Johann Karl Friedrich Zöllner, Charles Carleton Massey

Transcendental Physics

An Account of Experimental Investigations from the Scientific Treatises of Johann Carl Friedrich Zöllner

Johann Karl Friedrich Zöllner, Charles Carleton Massey

Transcendental Physics
An Account of Experimental Investigations from the Scientific Treatises of Johann Carl Friedrich Zöllner

ISBN/EAN: 9783337417000

Printed in Europe, USA, Canada, Australia, Japan

Cover: Foto ©berggeist007 / pixelio.de

More available books at **www.hansebooks.com**

TRANSCENDENTAL PHYSICS.

An Account of Experimental Investigations.

From the Scientific Treatises

OF

JOHANN CARL FRIEDRICH ZÖLLNER,

Professor of Physical Astronomy at the University of Leipsic;
Member of the Royal Saxon Society of Sciences;
Foreign Member of the Royal Astronomical Society of London;
of the Imperial Academy of Natural Philosophers at Moscow;
Honorary Member of the Physical Association at Frankfort-on-the-Main;
of the "Scientific Society of Psychological Studies," at Paris;
and of the "British National Association of Spiritualists," at London.

Translated from the German, with a Preface and Appendices, by

CHARLES CARLETON MASSEY,
OF LINCOLN'S INN, BARRISTER-AT-LAW.

LONDON:
W. H. HARRISON, 33 MUSEUM STREET, W.C.
1880.

CONTENTS.

	PAGE
TRANSLATOR'S PREFACE	xvii
AUTHOR'S DEDICATION TO MR. WILLIAM CROOKES, F.R.S. . .	xlv

CHAPTER I.

Gauss's and Kant's Theory of Space—The Practical Application of the Theory in Experiments with Henry Slade—True Knots produced upon a Cord with its ends in view and sealed together . 1

CHAPTER II.

Magnetic Experiments—Physical Phenomena—Slate-Writing under Test Conditions 22

CHAPTER III.

Permanent Impressions obtained of Hands and Feet—Proposed Chemical Experiment—Slade's Abnormal Vision—Impressions in a Closed Space—Enclosed Space of Three Dimensions open to Four-Dimensional Beings 48

CHAPTER IV.

Conditions of Investigation—Unscientific Men of Science—Slade's Answer to Professor Barrett 62

CHAPTER V.

Production of Knots in an Endless String—Further Experiments Materialisation of Hands—Disappearance and Reappearance of Solid Objects—A Table Vanishes, and afterwards Descends from the Ceiling in Full Light 71

CHAPTER VI.

Theoretical Considerations—Projected Experiments for Proof of the Fourth Dimension—The Unexpected in Nature and Life—Schopenhauer's "Transcendent Fate" . . . 93

CHAPTER VII.

Various Instances of the so-called Passage of Matter through Matter 103

CHAPTER VIII.

The Phenomena suitable for Scientific Research—Their Reproduction at Different Times and Places—Dr. Friese's and Professor Wagner's Experiments in Confirmation of the Author's . 130

CHAPTER IX.

Theoretical; "The Fourth Dimension"—Professor Hare's Experiments—Further Experiments of the Author with Slade—Coins Transferred from Closed and Fastened Boxes—Clairvoyance . 149

CHAPTER X.

An Experiment for Sceptics—A Wager—Slade's Scruples—A Rebuke by the Spirits—An Unexpected Result—Captious Objections 175

CHAPTER XI.

Writing through a Table—A Test in Slate-Writing conclusively Disproving Slade's Agency 191

CHAPTER XII.

A "Fault" in the Cable—A Jet of Water—Smoke—"Fire Everywhere"—Abnormal Shadows—Explanation upon the Hypothesis of the Fourth Dimension—A Séance in Dim Light—Movement of Objects—A Luminous Body 204

CHAPTER XIII.

Phenomena Described by Others 219

APPENDICES.

APPENDIX A.—The Value of Testimony in Matters Extraordinary 237
APPENDIX B.—Evidence of Samuel Bellachini, Court Conjurer at Berlin 259
APPENDIX C.—Admissions by John Nevil Maskelyne, and other Professional Conjurers . . 262
APPENDIX D.—Plate X. . . 265

LIST OF ILLUSTRATIONS.

—o—

                                                                PAGE

FRONTISPIECE.—The room at Leipsic in which most of the Experiments were conducted.

PLATE I.—Experiment with an Endless String . 15

,, II.—Leather Bands Interlinked and Knotted under Professor Zöllner's Hands 83

,, III.—Experiment with an Endless Bladder-Band and Wooden Rings 107

,, IV.—Result of the Experiment 109

,, V.—Ditto, on an Enlarged Scale . . 111

,, VI.—Experiment with Coins in a Secured Box 160, 161

,, VII.—The Representation of Conditions under which Slate-Writing was obtained 193

,. VIII.—Slate-Writing Extraordinary . 200, 201

,, IX.—Slate-Writing in five Different Languages 230, 231

,, X.—Details of the Experiment with an Endless Band and Wooden Rings 266

TRANSLATOR'S PREFACE.

> "These things, O Asclepius, will appear to be true if thou understandeth them, but if thou understandeth them not, incredible. For to understand is to believe, but not to believe is not to understand."—
> *The Divine Pimander.*

TRANSLATOR'S PREFACE.

---o---

"TRANSCENDENTAL PHYSICS" is the title of the third volume of Professor Zöllner's Scientific Treatises. Some of the parts comprised in the following translation belong to the earlier volumes, where the facts recorded are introduced in connection with the author's physical speculations. It is with some concern that the translator has been compelled to forego a full presentation of the latter; but it is hoped that enough is given to show their bearing upon facts, the public recognition of which is the principal object in view. With such assistance as is afforded by the author's occasional explanations, the English reader must be left to grapple, as best he can, with the unfamiliar conception of the fourth dimension of space. Professor Zöllner traces this hypothesis historically in the writings of some of the most eminent philosophers and mathematicians; but it was not possible to disengage this account from other metaphysical and scientific disquisitions, or from controversial topics, in which it is involved. A very general abstract is given in the first chapter, which is a reprint, by permission of Mr. Crookes, F.R.S., of an article published in the *Quarterly Journal of Science*, of April 1878, on the appearance of the first volume of the author's treatises.

The writer ventures to hope that this English version of facts so well-attested may be read by those to whom the intellectual worth and achievements of the principal witnesses are already known. But for the information of the general public, the following particulars concerning them are here given.

Professor Zöllner, the author and chief deponent, in whose house many of the facts he records occurred, was born in 1834, and is thus in the mature vigour of his intellectual life. He is Professor of Physics and Astronomy in the University of Leipsic, and has taken place in the front ranks of the scientific men of Europe. He has published many works, among which are *Sketches of a Universal Photometry of the Starry Heavens, Physical Nature of the Heavenly Bodies, The Nature of Comets*, and these treatises.

William Edward Weber, born 1804, is a Professor of Physics, and known as the founder, in common with his brother, of the doctrine of the Vibration of Forces. He has published an exhaustive work on *Electro-Dynamic Measurement* (4 vols. 1846–1854). No scientific reputation stands higher in Germany than that of Weber.

Professor Scheibner, of Leipsic University, is a well-known and highly distinguished mathematician.

Gustave Theodore Fechner, born 1801, is eminent

as a natural philosopher, and is likewise Professor of Physics at Leipsic. Among his works are *The Soul of Plants, The Zendavesta, The Things of the Future, Elements of Psycho-Physics, The Problem of the Soul,* and *About the Life Hereafter.*

It is not surprising that the testimony of these men, publicly given to such facts as those described in the following pages, has caused much excitement and controversy in Germany. The indisposition to see in the alleged phenomena of Spiritualism, as regards their reality and independence of known causes, a simple question of evidence, has been everywhere apparent. Nevertheless, it is just from this point of view that the public must, by degrees, be brought to regard the subject. The irrelevance of any other mode of treating it will sooner or later be recognised. The value of human testimony is determinable by known criteria, which can only be applied by a critical examination of the statements made, having regard also to what is ascertained about the witnesses. Supposing the veracity and intelligence of the latter to be above suspicion, we have to consider what were their opportunities for exact observation, with reference, of course, to the nature of the fact observed. The latter, indeed, is the main point, because we know that the faculty of accurate observation differs widely in different people, and we cannot have the same confidence in this special capacity of the witness that we may have in his general intelligence. For example: during the prosecution of Slade at Bow Street in 1876 by Professor Lankester, the latter declared himself unable to say on which side of a slate, pressed by Slade against the under surface of the table, a certain writing was produced; remarking, that the slate might have been reversed on withdrawal by sleight of hand, and that it was part of the art of a conjurer to effect such a change without observation. That there was some force in this suggestion could not be denied;* and had a witness in support of Slade stated that the writing, produced under the conditions described, appeared on the side of the slate which was pressed against the table, the accuracy of his observation, and therefore the value of his testimony, might have been challenged on this ground. How far a similar criticism is applicable to *any* of the facts here recorded by Professor Zöllner, to any of them, at least, which he lays stress upon, and thinks it worth while particularly to describe, the reader must judge for himself. But take, for instance, those of which circumstantial accounts are given at pages 14, 34, and 91, and contrast them, in relation to this all-important point, with the one just mentioned. If in the latter case, average powers of observation might possibly be baffled, in the others, the nature of the phenomena was such, that so far as Slade's physical instrumentality in them is in question, the suggestion that it

---

* The writer tested it afterwards, by making Slade withdraw the slate very slowly, inch by inch, as soon as the sound of writing ceased, when the writing appeared in successive lines on the *upper* surface of the slate (that against the table).

might have eluded observation will be felt to transgress the limits prescribed by candour and common sense.

The evidence of testimony is the evidence of the senses, or, to speak more accurately, of sense-impressions as interpreted by the mind, one degree removed. The only elements of fallacy possibly added by testimony to original observation are such as may result from defects of veracity, defects of memory, defects of judgment as to what is material to be mentioned, and defects of language, or the understanding of it by the recipient of the testimony. In short, the *peculiar* infirmity of proof by testimony is the uncertainty whether it conveys to the mind an exact or sufficient transcript of the fact as it was perceived by the original observer. For whatever concerns defects of observation belongs to the perception, and not to the transmission of facts by testimony. Even if we could be satisfied that we had got a perfect copy of the original objective impression made upon the witness, we still could not be sure that the fact so conveyed to us would not have contained more or less for ourselves had we been in his place. In all matters requiring skilled observation, it is obvious that the testimony of an expert has far higher value as evidence for those who are not experts than their own observations would have.

In considering what is the particular risk of any of the above-mentioned fallacies of testimony, except the first, attaching to a statement of facts, the same remark is applicable as has been already made in speaking of the value of original observations. Assuming the veracity of the witness, we must have especial regard to the *nature* of the facts, and of the statements concerning them. Are the latter so full, precise, and intelligible, as to evince that the witness is speaking from a strong recollection, with a clear appreciation of what is necessary to be known, and with a faculty of expression sufficient to convey distinct and unmistakable meanings to an ordinary understanding? In order to know whether the statements are really sufficient in these respects, we have to consider what is *the fact to be proved* from the particular facts described,—as in the following accounts the fact to be proved is, that Slade had no active physical participation in the production of what occurred. Just as the probative force of original observation depends on the supposition that no circumstance *which could have escaped observation* would impair the demonstration, so the probative force of testimony (veracity being granted) is additionally less in so far only as a similarly important circumstance, actually observed, could have escaped the memory, or be regarded in the judgment of the witness as too unimportant for mention. As the faculty of observation differs in different persons, so also differ the memory and intelligence. But as there are limits, which it would be contrary to all experience to suppose exceeded, to the fallibility of observation, so, likewise, we cannot ascribe to memory and intelligence, in reference to deliberate statements, defects so gross that we should infer with far more probability intentional untruth.

The validity of testimony to facts of such a nature, that *conceivable* errors of observation, memory, and judgment may be left out of account in concluding from them, is thus reduced to a question of veracity. There is no issue so studiously shirked as this by the people who heap contempt on all evidence in favour of occult phenomena. The imputation of lying is felt to be too crude, coarse, and unintelligent a way out of the difficulty. And so we hear a great deal of the folly and credulity of the witnesses, and of their unacquaintance with the wiles of the conjurer, but hardly at all of their mendacity. And yet no testimonies, taken singly at least, to such things are worth much, unless the issue can be narrowed to this of veracity. If there is any chance or possibility, consistently with the witnesses' truth, of the whole thing being only a more or less skilful trick, the testimony is scarcely worth adducing at all.

It will be seen that this demand, that trickery and conjuring shall not be suggested at large, without regard to the nature of what occurs, or to the conditions of its occurrence, as affording or not affording opportunities for the exercise of these arts, does not require from the critic of the evidence an exposition of the particular *modus operandi* of the supposed conjurer, which only the latter could supply. No one, for example, is called upon to explain the *modus operandi* of Psycho's performances at the Egyptian

Hall, or, as the alternative, to admit that some occult agency is concerned in them. It is enough that on Messrs. Maskelyne and Cooke's own stage, communication with the automaton, by means known to science, is possible, though Mr. Maskelyne stands at a distance from it, and no assistant is visible. But transport Mr. Maskelyne and his automaton to a private house, to a room which the conjurer has never entered before, or which he has had no facilities for adapting to his purposes; let him be unaccompanied by any assistant, and stand aside under close observation while Psycho plays his intelligent hand at whist with some of the company, or works out the sums in arithmetic which they set him. There would then arise a question for science the importance of which it could not affect to underrate. For on the assumption that Mr. Maskelyne was directly and consciously instrumental in what occurred, it must be that he had discovered a method of applying and directing natural forces at a distance,* while apparently himself inactive, which would be capable of the most practically important uses.†

If, now, the production of the true knots in an endless string, the rending of Professor Zöllner's bed-screen, the disappearance of the small table and its subsequent descent from the ceiling, in full light, in a private house, and under the observed conditions,

---

* That Psycho is not worked by Mr. Maskelyne himself from the stage by means of a magnet has been repeatedly demonstrated.
† See Appendix C.

of which the most noticeable is the apparent passivity of Slade during all these occurrences, are to be ascribed to any conscious operation of his, we can hardly avoid attributing to him scientific discoveries, or the possession of secrets of nature at least equally remarkable. But in that case he could, and it would clearly be his interest to produce these and similar astonishing effects with constant regularity. Professor Lankester would have witnessed them no less than Professor Zöllner, and Slade would long ago have amassed a fortune by his exhibitions. The fact that he cannot command these phenomena, at least the most striking of them, at will, points to conditions of their production varying with his own physical and mental states, and probably with those also of the persons resorting to him. And this is the reason why these phenomena, though as capable of verification by scientific men and trained observers (by whom they have in fact been repeatedly verified), as by any one else, are not exactly suitable for scientific verification. There is a clear distinction between the two things. Scientific verification supposes that the conditions of an experiment are ascertained, that they can be regularly provided, and the experiment repeated at pleasure. One hears occasionally of offers from men of science to investigate and attest certain phenomena of Spiritualism, selected by themselves, provided they can witness them under conditions of their own prescribing. These, in some cases well-meant overtures, proceed on the assumption that the phenomena, if genuine, require nothing but the mere physical presence of the medium, and that it is only necessary to take adequate precautions (no matter what these are) against deception by the latter, in order to obtain a scientific demonstration. When such an offer is rejected or neglected, the inference of course is drawn that the "phenomena" only occur when facilities for their fraudulent production are allowed. Yet it is equally consistent with the medium's knowledge that the conditions (of which he is himself ignorant) cannot be controlled, and with his consequent indisposition to be put upon a formal trial which may result in failure and discredit. Systematic investigation of this subject by men of science is much to be desired, but it must not be undertaken in a magisterial spirit, with the imposition of a test, and the demand of an immediate result. The only claim which spiritualists make upon scientists is that they shall not, in entire ignorance and contempt of the evidence, sanction and encourage the public prejudice by their authority.* But even this claim cannot be preferred with confidence. Since the Council of the Royal Society refused, by its rejection of Mr. Crookes's paper, " On the Experimental Investigation of a new Force," to be informed of the evidence, it must be considered that the Fellows of that distinguished body do, in general, dispose of the question on *a priori* grounds, and hold that no quantity or

---

* For example, by describing Spiritualism as "a kind of intellectual whoredom."—Professor Tyndall.

quality of human testimony can suffice to establish facts of this description, or even a *primâ facie* case in favour of them. So far as this peremptory rejection appeals to the principle of incredulity expounded in Hume's celebrated Essay on Miracles, it shows an utter ignorance of the reasoning by which that monstrous fallacy, and the contradictions in which its author involved himself, have been repeatedly exposed. This has never been better done than in the introduction to Mr. Alfred Russel Wallace's book, entitled "Miracles and Modern Spiritualism." There is, however, another proposition commonly put forward to dispose *a priori* of unacceptable testimony, substantially, but not logically, equivalent to Hume's, and which embodies a fallacy no less demonstrable, though so widely prevalent as to necessitate particular examination. This proposition is, that evidence, to command assent, should be proportional to the probability or improbability of the fact to be proved. Two years ago the writer dealt with it at some length in a paper read before the Psychological Society, and which is reprinted in an Appendix at the end of this volume.* Inasmuch as the fallacy in question, and the loose and inaccurate phrases, applied to the whole class of facts now in evidence, are in the nature of preliminary objections to the testimony about to be adduced, the reader is urgently referred to that essay, which cannot be conveniently comprised within the limits of a preface.

* Appendix A. "On the value of testimony in matters extraordinary."

Every opponent who recognises the obligation of dealing seriously with evidence, whether by explicitly objecting to its admissibility, or by questioning its intrinsic value, must be fairly and squarely encountered. To writers in the press who never miss an opportunity of discrediting Spiritualism, by derisive articles on the "exposures," real or reputed, of mediums, and on the occasional follies of spiritualists, only a passing word can be spared. To the present writer, at least, so-called Spiritualism represents no religious craze or sectarian belief, but an aggregation (not yet to be called a system) of proven facts of incalculable importance to science and speculation. Those who so regard the subject would be unmoved in their convictions of its truth and importance though it were proved that every medium was a rogue, and that many spiritualists were their willing dupes. Much of the evidence on which they rely has proceeded on that very assumption, and on the precautions which were accordingly taken. In none of it which is imparted to the public does the element of personal confidence in the medium enter in the smallest degree, though that feeling doubtless does and must often exist, especially when the manifestations occur, as they often do, in private families, and with persons whose characters are beyond all suspicion.

As regards the medium, Henry Slade, with whom Professor Zöllner's investigations were carried on, all the world knows, or did know a few years ago, that he was convicted at Bow Street Police Court, under the fourth section of the Vagrant Act, of using "subtle crafts and devices, by palmistry or otherwise," to deceive Professor E. Ray Lankester, F.R.S., and certain others; that he was sentenced by Mr. Flowers, the magistrate, to three months' imprisonment with hard labour; and that the conviction was afterwards quashed on appeal to the Middlesex Sessions, for a formal error in the conviction, as returned to that Court. Professor Zöllner gives the whole report of the various proceedings from beginning to end, at length, in his book, but it has not been thought necessary to reproduce it here. It may be stated generally that Professor Lankester had two sittings with Slade, at each of which he believed himself to have detected the mode in which the writing was produced on the slate. On the second occasion he was accompanied by a friend, Dr. Donkin, whose evidence agreed with his own. The *modus operandi*, according to these gentlemen, was this: Slade took one of his own slates, and held it for a time, concealed from the view of his visitors, between himself and the table, before placing it "in position," that is, pressed against the under surface of the corner of the table, for the pretended purpose of obtaining "Spirit-writing." During this interval the observers detected sounds as of writing, and observed motions of Slade's arm, suggestive that he was employed in writing on the slate, held, probably, between his legs. As to other "messages," obtained while the slate was in position, they supposed Slade to indite them by means of a bit of pencil stuck in the nail of one of his fingers. At length, after hearing writing as first described, Professor Lankester snatched the slate from Slade's hand as soon as it was placed against the table, and found the message already inscribed upon it.

Such was the clumsy trick—if trick indeed proceedings so imperfectly disguised can be called—of which a man who, if not a "medium," is unquestionably the most wonderful conjurer and illusionist in the world, was convicted, by inference, to use the magistrate's expression, "from the known course of nature." And there the matter might be left to the reader's reflections. Some few additional facts must, however, be stated. Previous to Professor Lankester's visit to him, Slade had been two months in London, being on his way to St. Petersburg, where he was under an engagement with a scientific committee of the Imperial University of that city. During this time he had been giving sittings to all comers, including not a few of literary and scientific attainments. We may safely conclude that the great impression he had produced was not the result of proceedings such as those described by his accusers. Among those named in the information against him, and whom he was charged, contrary, it was understood, to their express wish, with having deceived, were several well-known gentlemen, Dr. W. B. Carpenter, F.R.S., being one. Only one of these gentlemen, Mr R. H. Hutton, was called as a witness by the prosecution. His evidence was on the whole favourable to the accused. Of other witnesses called by the prosecution, not one professed to have detected trickery, though all seemed to suspect it. For the defence, it was proposed to call a number of witnesses of education and intelligence, for the purpose of giving evidence of phenomena—slate-writing and other—witnessed by them in Slade's presence, of a character and under conditions wholly inconsistent with any agency of his. Four only were allowed to give evidence, one of them being Mr. A. R. Wallace, the eminent naturalist. The present writer had been called by the prosecution (he being counsel in the case for another defendant), but believes that his evidence could not have been entirely satisfactory to that side. The effect of the evidence for the defence was described by the magistrate from the bench as "overwhelming;" but in giving judgment he expressly excluded it from consideration, confining himself to the evidence of the complainant, Professor Lankester, and of Dr. Donkin, and basing his decision upon "inferences to be drawn from the known course of nature"—a main question in the case being whether there are not some operations in nature *not* "generally known." An attempt had been made, with the wholly irregular assistance of Mr. John Nevil Maskelyne, the professional conjurer, to show that the table used by Slade, and which was produced in Court by the defence, was a "trick table," and expressly constructed to assist in the effects at the *séances*. This attempt utterly broke down. In order to allow room for the slate to be placed in the position usual for obtaining writing, a single central support was used for the flap of the table instead of side ledges. A wedge inserted at the pivot of this support had been pointed out as a most suspicious feature; it was explained by the carpenter that he had inserted it himself, without orders, for the simple purpose of remedying a defect in his own construction of the table. Professor Lankester, in his evidence, had described the table, before its production in Court, as without a frame, and as thus enabling Slade to move his legs and knees under it with greater facility. It turned out that the table had a frame of rather greater depth than usual. The table was impounded, and remained for several months in the custody of the Court, and open to inspection and examination for concealed magnets, and so forth. None were discovered, and the table is now at the rooms of the British National Association of Spiritualists, at 38 Great Russell Street, where it can be seen by the curious.

Nothing was more prejudicial to Slade, or more tended to produce the impression that he was an impostor than his ascribing the "messages" on his slates to spirits of the dead. "Allie," his deceased wife; Professor Lankester's fictitious "Uncle John:" the random names that came, and the messages of recognition to which they were signed, naturally seemed to the public, little accustomed, or, in this case, disposed to distinguish issues, even more indicative of fraud than the direct evidence. The writer, from the intimate knowledge he acquired of Slade, is satisfied that the latter really believed in the identity of his "spirits." Nor was this belief at all unnatural. A large proportion of his visitors do obtain writing signed by the names of deceased friends of whom usually Slade has never heard; this being often the case with strangers visiting him for the first time. That there is any "pumping" process applied to his visitors before sitting for the writing is utterly untrue. This suggestion was put forward as part of the case of the prosecution in the opening statement; and it had its effect on the public mind; but *not one particle* of evidence was adduced in support of it; on the contrary, all the witnesses, upon cross-examination, admitted that no questions were put to them, nor was any attempt made to draw them into conversation before the sittings; and it was on this ground that a charge of conspiracy against Slade and another defendant, Simmons, broke down and was dismissed. It was a suggestion which seems to have been made simply because, on the assumption that Slade was an habitual impostor, *it ought to have been true*, and perhaps it was expected that something of the sort would turn up in the evidence.

To the writer, it has always appeared that the presence of a departed friend, *in propriâ personâ*, is very insufficiently proved by communications purporting to be thus derived, even when all knowledge by the medium of the name of the deceased, or of the circumstances called to the recollection of the survivor by way of identification, can be conclusively disproved. We are so profoundly ignorant of the deeper mysteries of life, that in this region we are not entitled to accept an explanation as true simply because it is sufficient, and because we cannot represent to ourselves any other. Usually, in the writer's experience invariably, in these communications any attempt to pursue the test by further probing the memory and intelligence of the supposed spirit results in failure. And the frequency of admittedly deceptive communications proves at least that there are mixed influences abroad, and that the hospitality of the medium's spiritual neighbourhood is shared by very questionable guests. Some time before the commencement of the proceedings against Slade, the writer, being extremely sceptical of spirit-identity, wrote a fictitious name on the back of a slate (carefully concealing the side on which he wrote, and the motions of the pencil), and handing the slate, clean side uppermost, to Slade, requested that the individual whose name was written would communicate, if present. Slade took the slate without reversing it, and laid a morsel of pencil upon it; then *at once* pressed it against the under surface of the corner of the table, so that the clean side was in contact with that surface, the side on which the name was written being the lower one. Writing was heard directly, and the slate being withdrawn and immediately inspected, on its *upper* side was found a kind little message of friendly remembrance signed by the fictitious name. Never was the writer more satisfied of Slade's integrity than on this occasion, and the circumstance is only mentioned here to show how distinct are issues which were confused in the Slade prosecution. Such experiments, however, are regarded by spiritualists as highly objectionable. They believe, and they have some grounds in experience for their belief, that fraud in the investigator will, by a subtle attraction, elicit fraud in the manifestations. Some go further, and maintain that the strong *animus* of prejudice, unconsciously but powerfully willing the very appearances it expects, may mesmerically control the sensitive medium, and force his actions in the direction it dictates.

But to return from this digression :

Immediately after the conviction was quashed, Professor Lankester applied for and obtained a fresh summons against Slade, as it was stated, "in the interests of science." (He had already, in the " Times," described the proceedings of the British Association as having been "degraded" by the introduction of the subject of Spiritualism, on which Professor Barrett had read a paper.) But meanwhile, Slade had broken down under the pressure of anxiety, and the agitation caused by public contumely and his own indignant sense of wrong. He had resolutely refused to listen to suggestions that he should leave the country, by consent of his bail, before the appeal case came on. As the time approached, he had a slight attack of brain-fever, as was certified by two physicians. During its continuance he was occasionally delirious, and the writer saw him in this condition. Partially recovered, he with difficulty dragged himself to the Court; he appeared apathetic and almost unconscious during its critical proceedings. It was the belief of his friends that further persecution would kill him outright; but independently of this, immediate change of scene and associations was imperatively necessary to his recovery. He left England with his niece and with his friend Mr. Simmons a day or two after the appeal case was determined. From The Hague, after a rest of a few months, he addressed, through Mr. Simmons, the following offer to his accuser:—

"Professor E. R. LANKESTER—Dear Sir,—Dr. Slade having in some measure recovered from his very severe illness, and his engagement to St. Petersburg having been postponed (by desire of his friends there) till the autumn, desires me to make the following offer:—

"He is willing to return to London for the express and sole purpose of satisfying you that the slate-writing occurring in his presence is in no way produced by any trickery of his. For this purpose he will come to your house unaccompanied by any one, and will sit with you at your own table, using your own slate and pencil; or, if you prefer to come to his room, it will suit him as well.

"In the event of any arrangement being agreed upon, Slade would prefer that the matter should be kept strictly private.

"As he never can guarantee results, you shall give him as many as six trials, and more if it shall be deemed advisable. And you shall be put to no charge or expense whatever.

"You on your part shall undertake that during the period of the sittings, and for one week afterwards, you will neither take nor cause to be taken, nor countenance legal proceedings against him or me. That if in the end you are satisfied that the slate-writing is produced otherwise than by trickery, you shall abstain altogether from further proceedings against us, and suffer us to remain in England, if we choose to do so, unmolested by you.

"If, on the other hand, you are not so satisfied, you shall be at liberty to proceed against us, after the expiration of one week from the conclusion of the six or more experiments, if we are still in England. You will observe that Slade is willing to go to you without witnesses of his own, and to trust entirely to your honour and good faith.

"Conscious of his own innocence, he has no malice against you for the past. He believes that you were very naturally deceived by appearances which to one who had not previously verified the phenomena under more satisfactory conditions may well have seemed suspicious. Should we not hear from you within ten days from this date, Slade will conclude that you have declined his offer.—I have the honour to be, Sir, your obedient servant, J. SIMMONS."

"37 SPUI STREET, THE HAGUE, *May 7th*, 1877.

To this letter no answer was ever received.

After a long rest on the Continent, Slade was able to give the wonderful *séances* recorded in this volume. He went on to St. Petersburg and fulfilled his engagement there. Returning to London for a day or two in 1878, he embarked for Australia, and made a great impression in the colonies. He returned to America by San Francisco last year, and is now once more in New York. During his travels after leaving England, he is said to have suffered from a partial paralysis, induced by his troubles here.

With Slade, as with no other medium known to the writer, the conditions of investigation are essentially simplified by the fact that he invariably sits with his visitors in a full light. "In the interests of science" it is greatly to be desired that he may be able to revisit London, liberated by an improved state of public opinion from all danger of molestation. We who urge the truth of these things are only anxious that the investigation should be conducted in the light of day, and by the most competent persons. So strong is this feeling, that it is believed a fund would easily be raised for the purpose of bringing Slade over to England and placing him in the hands of a scientific committee who should examine this question of the slate-writing with the facilities suggested in the offer of Slade to Professor Lankester. That the Slade prosecution was designed to deal a blow at Spiritualism, or rather at the serious investigation of facts which are usually included in that term, will hardly be doubted. But without in the least questioning that Professor Lankester had in his own belief, as in that of the vast majority of the public, the strongest justification for the course he took, it is to be trusted that a truer appreciation of the interests of science will shortly prevail. Professor Zöllner in these volumes, speaking from the point of view of a true man of science, expresses his indignation at these transactions in England, and at the unmeasured abuse of Slade in the German press, in strong terms. The translator has thought it better to omit all this, leaving the facts to speak for themselves, and in the assurance that hereafter, if not at present, public opinion will pronounce a just judgment upon them.

Professor Zöllner's polemic, referred to in his dedication to Mr. Crookes, has a far wider scope and application than will be apparent from the following translation. He has set himself, in the course of these treatises, to encounter with unsparing force certain tendencies among men of science, and in the Press, which he regards as demoralising in the highest degree. All particular reference to these subjects is here avoided. This is almost exclusively a volume of evidences, and the introduction of other topics of controversy might not be favourable to the judicial calmness with which the former should be considered. Nevertheless, the belief may be avowed that the substantiation of the facts before us, in scientific and public opinion, cannot fail to have, indirectly, a revolutionary effect on many departments of speculation and practice. All that is asked at present, however, is a fair judgment on the facts themselves, without regard to the possible extent of their significance. For further, and very striking evidences of the phenomenon of writing by unknown agencies, the reader is referred to a small volume entitled "Psychography," by M. A. (Oxon): (Harrison, London, 1878).

Although the popular suggestion, that the phenomena of Spiritualism are merely conjuring under false pretences, will not find acceptance with any one who seriously considers the evidence, it has been thought worth while to meet it additionally by the testimonies of some well-known experts in the arts of illusion. These will be found in Appendices B and C; the evidence of Bellachini, Court conjurer at Berlin, who was employed to conduct a systematic investigation of the phenomena in Slade's presence, is especially remarkable.

The literary merit of the following translation is of such infinite unimportance in comparison with the matter that the writer hardly cares to disarm criticism on this point, provided the substantial accuracy of the rendering is not impugned. He is quite sensible of its other defects, and has only to plead that he is almost entirely self-taught in German, having never visited countries in which it is spoken, or studied for any length of time under a master. He has only undertaken the work because it seemed that otherwise it would not be done at all, or at least not yet. Nor has he any pecuniary interest in it. He now gives it to the English public, in the hope that it may conduce to a more rational appreciation and to a juster treatment of evidence on this subject than has hitherto prevailed.

AUTHOR'S DEDICATION.

AUTHOR'S DEDICATION.

To WILLIAM CROOKES, F.R.S.

WITH the feeling of sincere gratitude, and recognition of your immortal deserts in the foundation of a new science, I dedicate to you, highly honoured colleague, this Third Volume of my Scientific Treatises. By a strange conjunction our scientific endeavours have met upon the same field of light, and of a new class of physical phenomena which proclaim to astonished mankind, with assurance no longer doubtful, the existence of another material and intelligent world. As two solitary wanderers on high mountains joyfully greet one another at their encounter, when passing storm and clouds veil the summit to which they aspire, so I rejoice to have met you, undismayed champion, upon this new province of science. To you, also, ingratitude and scorn have been abundantly dealt out by the blind representatives of modern science, and by the multitude befooled through their erroneous teaching. May you be consoled by the consciousness that the undying splendour with which the names of a Newton and a Faraday have illustrated the history of the English people can be obscured by nothing, not even by the political decline of this great nation: even so will your name survive in the history of culture, adding a new ornament to those with which the English nation has endowed the human race. Your courage, your admirable acuteness in experiment, and your incomparable perseverance, will raise for you a memorial in the hearts of grateful posterity, as indestructible as the marble of the statues at Westminster. Accept, then, this work as a token of thanks and sympathy poured out to you from an honest German heart. If ever the ideal of a general peace on this earth shall be realised, this will assuredly be the result not of political speeches and agitations, in which human vanity always demands its tribute, but of the bond of extended knowledge and advancing information, for which we have to thank such heroes of true science as Copernicus, Galileo, Kepler, Newton, Faraday, Wilhelm Weber, and yourself.* * * *

In the first place it is necessary that the truth should be regardlessly outspoken, in order to encounter lies and tyranny, no matter under what shape they threaten to impede human progress, with energy and effect. In this sense I beg you to judge my combat against scientific and moral offences, not only in my own, but also in your country.

---

* Here follow references to subjects of controversy foreign to the purpose of this translation, and which occupy much of this, as of the preceding volumes of the treatises.

Every polemic, even the justest, has in it something uncongenial, like the sight of a battle or of a bloody battle-field. For hereby is man reminded impressively of the imperfections and faults of his earthly existence. And yet are gathered the noblest blossoms of the human heart, in its self-renouncing devotion of the dearest to the Fatherland, round the graves of the fallen warriors. The poetry and history of all peoples glorify these blood-saturated spots with their noblest breath, and the returning spring sees crosses woven with roses and ivy, where a year before the battle raged. So, hereafter, will this literary battle-field appear to the generation growing up. They will have understood the moral necessity of the strife, and in the morning splendour of a new epoch of human culture will have forgotten the repulsiveness (*das Unsympathische*) of my polemic.

But united England and Germany may always remember the words of your great physicist, Sir David Brewster, who, in his "Life of Newton," reminds us of the indestructibility and immortality of the works of human genius:—

"The achievements of genius, like the source from which they spring, are indestructible. Acts of legislation and deeds of war may confer a high celebrity, but the reputation which they bring is only local and transient; and while they are hailed by the nation which they benefit, they are reproached by the people whom they ruin or enslave. The labours of science, on the contrary, bear along with them no counterpart of evil. They are the liberal bequests of great minds to every individual of their race, and wherever they are welcomed and honoured, they become the solace of private life, and the ornament and bulwark of the commonwealth."

With these consolatory words of one of your celebrated countrymen, accept, my honoured friend, the present work as a token of the sincere esteem of the Author.

LEIPSIC, *October 1st*, 1879.

TRANSCENDENTAL PHYSICS.

Chapter First.

ON SPACE OF FOUR DIMENSIONS.*

GAUSS'S AND KANT'S THEORY OF SPACE—THE PRACTICAL APPLICATION OF THE THEORY IN EXPERIMENTS WITH HENRY SLADE—TRUE KNOTS PRODUCED UPON A CORD WITH ITS ENDS IN VIEW AND SEALED TOGETHER.

IN the first treatise the author shows that both Newton and Faraday were advocates of the theory of

* This first chapter consists of an article which appeared in the *Quarterly Journal of Science*, April 1878, and is reprinted here by permission of Mr. William Crookes, F.R.S. The facts are from "Wissenschaftliche Abhandlungen von Joh. Carl Friedrich Zöllner, Professor der Astrophysik an der Universität zu Leipzig. Erster Band. Leipzig: L. Staackmann, 1878. (With portraits and facsimiles of Newton, Kant, and Faraday. 8vo, 732 pages.)"

CONTENTS.

1. On Action at a Distance.
2. Emil du Bois Reymond and the Limits of Natural Knowledge.
3. Newton's Law of Gravitation, and its Derivation from the Static Effects of Electricity.
4. The Laws of Friction, and their Deduction from the Dynamic Effects of Electricity.
5. As to the Existence of Moving Electric Particles in *All* Bodies.
6. Adhesion and Cohesion as deducted from the Dynamic Forces of Electricity.
7, 8, 9. The Mechanical, the Magnetical, and the Electrical Effects of Light and of Radiant Heat.
10. Radiometrical Researches.
11, 12. On the Theory of Electric Emission and its Cosmical Application.
13. Thomson's Demons and the Phantoms of Plato.

A

direct action at a distance through a vacuum, in opposition to the views of many modern scientific men. In the last treatise, which is of the highest interest, the author describes experiments which he made in Leipzig, in December 1877, with Mr. Henry Slade, the American. These experiments were only the practical application of Gauss's and Kant's theory of space, which these two eminent men imagined might contain more than three dimensions. The author will try to give to the readers of the *Quarterly Journal of Science* an idea of this theory, though he must of course refer to the work itself for a more ample explanation of it.

In accordance with Kant, Schopenhauer, and Helmholtz, the author regards the application of the law of causality as a function of the human intellect given to man *à priori*, *i.e.*, before all experience. The totality of all empirical experience is communicated to the intellect by the senses, *i.e.*, by organs which communicate to the mind all the sensual impressions which are received at the *surface* of our bodies. These impressions are a reality to us, and their sphere is two-dimensional, acting not in our body, but only on its *surface*.

We have only attained the conception of a world of objects with three dimensions by an intellectual process. What circumstances, we may ask, have compelled our intellect to come to this result? If a child contemplate its hand, it is conscious of its existence in a double manner—in the first place by its tangibility, in the second by its image on the retina of the eye. By repeated groping about and touching, the child knows by experience that his hand retains the same form and extension through all the variations of distance and positions under which it is observed; notwithstanding that the form and extension of the image on the retina constantly change with the different position and distance of the hand in respect to the eye. The problem is thus set to the child's understanding, How to reconcile to its comprehension the apparently contradictory facts of the *invariableness* of the object, together with the *variableness* of its appearance. This is only possible within space of three dimensions, in which, owing to perspective distortions and changes, these variations of projection can be reconciled with the constancy of the form of a body.

So, likewise, in the stereoscope, the representation of the corporeality—*i.e.*, of the third dimension—springs up in our mind when the task is presented to our intellect to refer at once two different plane pictures, without contradiction, to one single object.

Consequently our contemplation of a three-dimensioned space has been developed by means of the law of causality, which has been implanted in us *à priori*, and we have come to the idea of the third dimension in order to overcome the apparent inconsistency of facts, of the existence of which experience daily convinces us.

The moment we observe in three-dimensioned space contradictory facts, *i.e.*, facts which would force us to ascribe to a body two attributes or qualities which hitherto we thought could not exist together,—the moment, I say, in which we should observe such contradictory facts in a three-dimensioned body, our reason would at once be forced to reconcile these contradictions.

There would be such a contradiction, for example, if we were to ascribe to one and the same object at once mutability and immutability, the most universal attribute of a body being the quantity of its ponderable matter. In conformity with our present experience we consider this attribute as unalterable. As soon, however, as phenomena occur which prove it to be alterable, we shall be obliged to generalise our representation of the ideality of a body so as to bring the observed change in the quantity of its matter in accordance with its hitherto-imagined unchangeableness.

On page 235 of his book the author quotes the celebrated mathematician Riemann, who says in his work *Concerning the Hypotheses upon which Geometry is founded:*—

"The explanation of these facts can only be found by starting from the actual theories of the appearance of all phenomena which are confirmed by experience, and of which, as they now are, Newton has laid the foundation. Urged forward by facts, which we cannot explain through our hitherto-conceived theories, we slowly remodel our conceptions. If phenomena occur which, according to our conception, were to be expected with probability, our theories are confirmed, and our confidence in them is founded upon this confirmation by experience. If, however, something occurs which we do not expect, which according to our theory was improbable or impossible, the task is imposed on us to remodel our theory, in order to make the observed facts cease to be in contradiction with our improved theory. The completion of our system of ideas forms the explanation of the unexpected observation. Our conception of nature by this process grows slowly to be more complete and more just, at the same time it retreats more and more beneath the surface of appearances."

I now proceed to apply the higher conception of space to the theory of twisting a perfectly flexible cord. Let us consider such a cord to be represented by $a\ b$, showing us, when stretched, a development of space in *one* dimension—

If the cord is bent so that during this action its parts always remain in the same plane, a development of space in *two* dimensions will be required for this operation. The following figure may be given to the cord :—

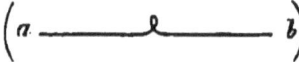

and all its parts, if conceived of infinite thinness, may be considered as lying in the same plane, *i.e.*, in a development of space in two dimensions. If the flexible cord, without being broken, has to be brought back into the former figure of a straight line, in such a manner that during this operation all its parts remain in the same plane, this can only be effected by describing with one end of the cord a circle of 360°.

For beings with only *two*-dimensional perceptions these operations with the cord would correspond to what we, with our *three*-dimensional perception, call a knot to the cord. Now if a being, limited on account of its bodily organisation to the conception of only *two* dimensions of space, possessed, nevertheless, the ability of executing by his will operations with this cord which are only possible in the space of *three* dimensions, such a being would be able to undo this two-dimensional knot in a much simpler way. Merely the turning over of part of the cord would be required, so that after the operation, when all parts again lie in the same plane, the cord would have passed through the following positions :—

By the same operations, but in an inverted sense, such a being would be able again to form the knot without needing that circumstantial process, during which all parts of the thread have to remain in the *two*-dimensional space of perception.

If this consideration, by way of analogy, is transferred to a knot in space of *three* dimensions, it will easily be seen that the tying as well as the untying of such a knot can only be effected by operations, during which the parts of the cord describe a line of *double* curvature, as shown by this figure—

We three-dimensional beings can only tie or untie such a knot by moving one end of the cord through 360° in a plane which is *inclined* towards that other plane containing the two-dimensional part of the knot. But if there were beings among us who were able to produce by their will four-dimensional movements of material substances, they could tie and untie such knots in a much simpler manner by an operation analogous to that described in relation to a two-dimensional knot.

It is by no means necessary — nay, not even *probable*—that such beings should have a contemplative consciousness of these actions of their will. For all our conceptions in relation to the movements of our limbs, and to those produced by their means in other bodies, have been acquired by us solely by way of *experience*. Having observed from childhood that a voluntary movement of our limbs is always connected with a corresponding change in our visional impressions, accompanying the action of our will, it is only in this way that we are now able to connect the movements of our body or of other objects with a corresponding conception of such motion.

Berkeley demonstrated this truth in the year 1709, in his *Essay Towards a new Theory of Vision* and in his *Principles of Human Knowledge*. In the last-mentioned treatise he remarks, on the relation of our visional perceptions to the sensations of touch :—

> "So that in strict truth the ideas of sight, when we apprehend by them distance, and things placed at a distance, do not suggest or mark out to us things actually existing at a dis-

tance, but only admonish us what ideas of touch will be imprinted in our minds at such and such distance of time, and in consequence of such or such actions."—Berkeley, *Principles of Human Knowledge* (Fraser's Edition, vol. i. p. 177).

Lichtenberg, in 1799, expresses himself in like manner when he says:—

"To perceive something *outside* ourselves is a contradiction; we perceive only *within* us; that which we perceive is merely a modification of ourselves, therefore, *within* us. Because these modifications are independent of ourselves, we seek their cause in *other* things that are outside, and say there are *things beyond us*. We ought to say '*præter nos;*' but for '*præter*' we substitute the preposition '*extra*,' which is something quite different, i.e., we *imagine* these things in the space *outside* ourselves. This evidently is *not* perception, but it seems to be something firmly interwoven with the nature of our sensual perceptive powers; it is the form under which that conception of the '*præter nos*' is given to us—the form of the sensual."

The want of these conceptions would necessarily be felt by us, if in some individuals, and these only occasionally, the will should be capable of producing physical movements, for whose geometro-mathematical definition a four-dimensional system of co-ordinates is necessary.

To my knowledge Gauss was the first to direct, from the point of view of the "Geometria Situs," his attention to the theory of the twistings of flexible cords. In his manuscripts left behind (Gauss's Werke, vol. v. p. 605) we find the following remarks:—

"Of the *Geometria Situs* which Leibnitz foresaw, and on which to throw a feeble glance was allowed only to a few mathematicians (Euler and Vandermonde), we, after a lapse of 150 years, know and possess hardly more than nothing. One of the principal problems on the boundary of the *Geometria Situs* and the *Geometria Magnitudinis* will be to calculate the number of the twistings of two closed and endless cords."

In my first treatise, *On Action at a Distance*, I have discussed in detail the truth, first discovered by Kant, later by Gauss and the representatives of the anti-Euclidian geometry, viz., that our present conception of space, familiar to us by habit, has been derived from experience, *i.e.*, from empirical facts by means of the causal principle existing *à priori* in our intellect. This in particular is to be said of the three dimensions of our present conception of space. If from our childhood phenomena had been of daily occurrence, requiring a space of four or more dimensions for an explanation which should be free from contradiction, *i.e.*, conformable to reason, we should be able to form a conception of space of four or more dimensions. It follows that the *real* existence of a four-dimensional space can only be decided by *experience*, *i.e.*, by observation of *facts*.

A great step has been made by acknowledging that the *possibility* of a four-dimensional development of space can be understood by our intellect, although, on account of reasons previously given, no corresponding image of it can be conceived by the mind. (Dass die moeglichkeit eines vierdimensionalen Raumgebietes *begrifflich* ohne Widerspruch *denkbar*, wenn auch nicht *anschaulich vorstellbar* ist.)

But Kant advances one step farther. From the *logically* recognised *possibility* of the existence of space having more than three dimensions, he infers their "very probably *real* existence" when he verbally remarks :—

"If it is *possible* that there be developments of *other* dimensions in space, it is also very *probable* that God has somewhere produced them. For His works have all the grandeur and variety that can possibly be comprised."

"In the foregoing I have shown that several worlds, taken in a metaphysical sense, *might* exist together, but, at the same time, here is the *condition*, which, according to my belief, is the only one which makes it probable that several such worlds *really exist*."—(Kant's Works, vol. v. p. 25.)

I may further cite the following observations of Kant:—

"I confess I am much inclined to assert the existence of immaterial beings in this world, and to class my soul itself in the category of these beings."

"We can imagine the possibility of the existence of immaterial beings without the fear of being refuted, though, at the same time, without the hope of being able to demonstrate their existence by reason. Such spiritual beings would exist in space, and the latter notwithstanding would remain penetrable for material beings, because their presence would imply an acting power in space, but not a *filling* of it, *i.e.*, a resistance causing solidity."

"It is, therefore, as good as demonstrated, or it could easily be proved, if we were to enter into it at some length; or, better still, *it will be proved in the future—I do not know where and when—that also in this life the human soul stands in an indissoluble communion with all the immaterial beings of the spiritual world; that it produces effects in them, and in exchange receives impressions from them, without, however, becoming humanly conscious of them, so long as all stands well."*

"It would be a blessing if such a systematic constitution of the spiritual world, as conceived by us, had not merely to be inferred from the —too hypothetical—conception of the spiritual nature generally, but would be inferred, or at least conjectured, as probable from some *real and generally acknowledged observation."* —(Kant's Works, vol. vii. p. 32.)

I have already in the above-cited treatise discussed some physical phenomena, which must be possible for such four-dimensional beings, provided that under certain circumstances they are enabled to produce effects in the real material world that would be visible, *i.e.*, conceivable to us three-dimensional beings. As one of these effects, I discussed at some length the knotting of a single endless cord. If a single cord has its ends tied together and sealed, an intelligent being, having the power voluntarily to produce on this cord four-dimensional bendings and movements, must be able, *without* loosening the seal, to tie one or more knots in this endless cord.

Now, this experiment has been successfully made within the space of a few minutes in Leipzig, on the 17th of December 1877, at 11 o'clock A.M., in the presence of Mr. Henry Slade, the American.

The accompanying engraving (Plate I.) shows the strong cord with the four knots* in it, as well as the position of my hands, to which Mr. Slade's left hand and that of another gentleman were joined. While the seal always remained in our sight on the table, the unknotted cord was firmly pressed by my two thumbs against the table's surface, and the remainder of the cord hung down in my lap. I had desired the tying of only *one* knot, yet the *four* knots—minutely represented on the drawing—were formed, after a few minutes, in the cord.

The hempen cord had a thickness of about 1 millimètre; it was strong and new, having been bought by myself. Its single length, before the tying of the knots, was about 148 centimetres; the length therefore of the doubled string, the ends having been joined, about 74 centims. The ends were tied together in an ordinary knot, and then—protruding from the knot by about 1.5 centims.—were laid on a piece of paper and sealed to the same with ordinary sealing-wax, so that the knot just remained visible at the border of the seal. The paper round the seal was then cut off, as shown in the illustration.

The above described sealing of two such strings, with *my own* seal, was effected *by myself* in my apartments, on the evening of December 16th, 1877,

* In the enlarged drawings the knots have been represented by mistake symmetrical; they were tied on one side, in accordance with the small figure of the cord.

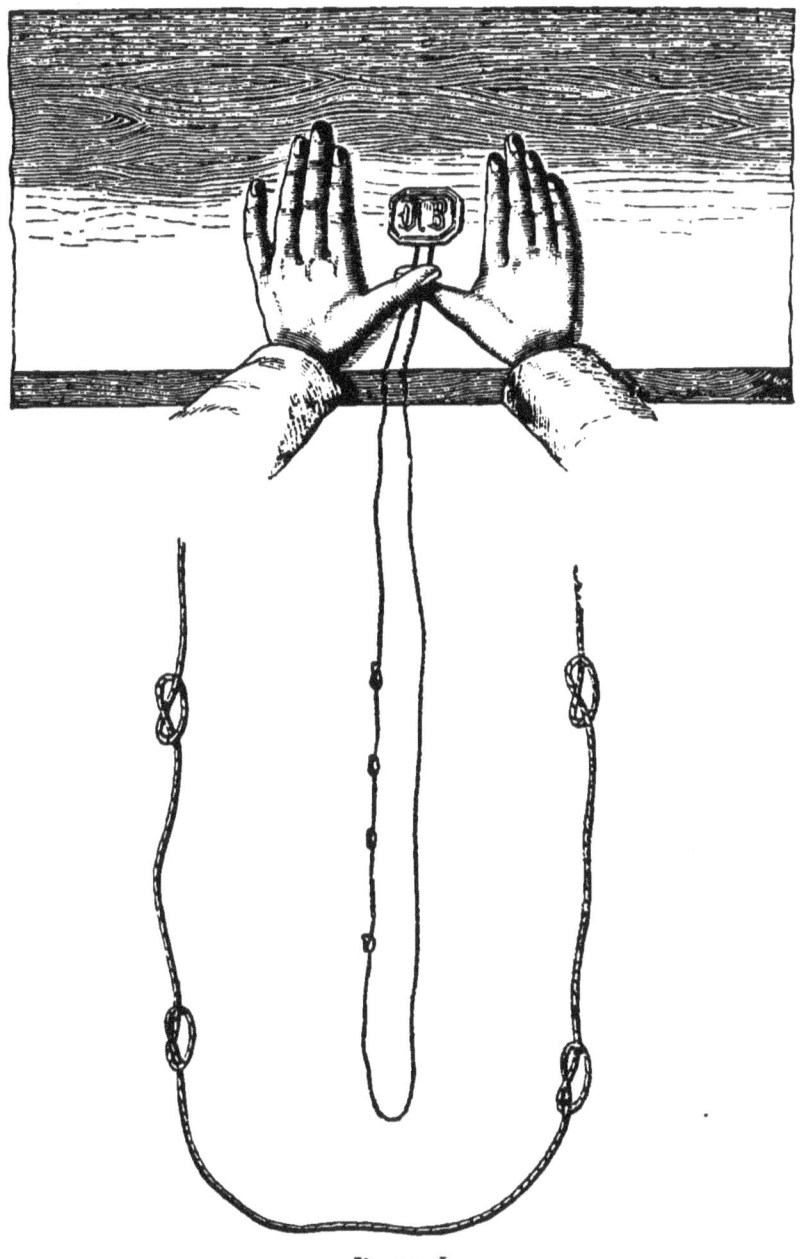

PLATE I.

at 9 o'clock, under the eyes of several of my friends and colleagues, and *not* in the presence of Mr. Slade. Two other strings of the same quality and dimensions were sealed by Wilhelm Weber with *his* seal, and in his own rooms, on the morning of the 17th of December, at 10.30 a.m. With these four cords I went to the neighbouring dwelling of one of my friends, who had offered to Mr. Henry Slade the hospitalities of his house, so as to place him exclusively at my own and my friend's disposition, and for the time withdrawing him from the public. The *séance* in question took place in my friend's sitting-room immediately after my arrival. I myself selected one of the four sealed cords, and, in order never to lose sight of it before we sat down at the table, I hung it around my neck—the seal in front always within my sight. During the *séance*, as previously stated, I constantly kept the seal — remaining unaltered — before me on the table. Mr. Slade's hands remained *all the time* in sight ; with the left he often touched his forehead, complaining of painful sensations. The portion of the string hanging down rested on my lap,—out of my sight, it is true,—but Mr. Slade's hands *always* remained visible to me. I particularly noticed that Mr. Slade's hands were not withdrawn or changed in position. He himself appeared to be perfectly passive, so that we cannot advance the assertion of his having tied those knots by his *con-*

scious will, but only that they, under these detailed circumstances, were formed in *his presence* without *visible* contact, and in a room illuminated by bright daylight.

According to the reports so far published, the above experiment seems also to have succeeded in Vienna in presence of Mr. Slade, although under less stringent conditions.* Those of my readers who wish for further information on other physical phenomena which have taken place in Mr. Slade's presence, I refer to these two books. I reserve to later publication, in my own treatises, the description of further experiments obtained by me in twelve *séances* with Mr. Slade, and, as I am expressly authorised to mention, in the presence of my friends and colleagues, Professor Fechner, Professor Wilhelm Weber, the celebrated electrician from Göttingen, and Herr Scheibner, Professor of Mathematics in the University of Leipzig, who are *perfectly* convinced of the reality of the observed facts, altogether excluding imposture or prestidigitation.

At the end of my first treatise, already finished in manuscript in the course of August 1877, I called attention to the circumstance that a certain number of physical phenomena, which, by "synthetical con-

* Mr. Slade's "Aufenthalt in Wien: Ein offener Brief an meine Freunde." Wien: I. C. Fischer & Co., 1878. "Der Individualismus in Lichte der Biologie und Philosophie der Gegenwart von Lazar B. Hellenbach." Wien: Braumüller, 1878.

clusions *à priori*," might be explained through the generalised conception of space and the platonic hypothesis of projection, coincided with so-called spiritualistic phenomena. Cautiously, however, I said :—

"To those of my readers who are inclined to see in spiritualistic phenomena an *empirical* confirmation of those phenomena above deduced in regard to their *theoretical* possibility, I beg to observe that from the point of view of idealism there must first be given a precise definition and criticism of *objective reality*. Indeed, if *everything* perceivable is a conception produced in us by *unknown* causes, the distinguishing characteristic of the *objective* reality from the *subjective* reality (phantasma) cannot be sought in nature, but only in accidental attributes of that process, producing conceptions. If causes unknown to us produce simultaneously in several individuals the same conception, only subject to those distinctions which depend upon differences in the position of the observers, we refer such conception to a *real* object *outside* of us; this conception not taking place, we refer that conception to causes *within* us, and call it hallucination.

"Now, whether the spiritualistic phenomena

belong to the first or to the second category of these conceptions, I do not venture to decide, so far never having witnessed such phenomena. On the other hand, I do not possess, with regard to men like Crookes, Wallace, and others, such an exalted opinion of *my own* intellect, as to believe that I myself, under similar conditions, should not be subject to the same impressions." (Written in August 1877.)

This supposition received, four months after my writing it down, a full confirmation by the above-mentioned experiments with the American, Mr. Henry Slade. In making them I was intent upon giving full consideration to the above-cited distinction between a subjective phantasma and an objective fact. The four knots in the above-mentioned cord, with the seal unbroken, this day still lie before me; I can send this cord to any man for examination; I might send it by turn to all the learned societies of the world, so as to convince them that not a *subjective* phantasma is here in question, but an *objective* and lasting effect produced in the material world, which no human intelligence, with the conceptions of space so far current, is able to explain.

If, nevertheless, the foundation of this fact, deduced by me on the ground of an enlarged conception of space, should be denied, only one other kind of explanation would remain, arising from a moral mode of consideration that at present, it is true, is quite customary. This explanation would consist in the presumption that I myself and the honourable men and citizens of Leipzig, in whose presence several of these cords were sealed, were either common impostors, or were not in possession of our sound senses sufficient to perceive if Mr. Slade himself, before the cords were sealed, had tied them in knots. The discussion, however, of such a hypothesis would no longer belong to the dominion of science, but would fall under the category of social decency.

Some other still more surprising experiments—prepared by me with a view to further testing this theory of space—have succeeded, though Mr. Slade thought their success impossible. The sympathising and intelligent reader will be able to understand my delight caused thereby. Mr. Slade produced on me and on my friends the impression of his being a gentleman : the sentence for imposture pronounced against him in London necessarily excited our *moral* sympathy, for the *physical* facts observed by us in so astonishing a variety in his presence, negatived on every reasonable ground the supposition that he in one solitary case had taken refuge in wilful imposture. Mr. Slade, in our eyes, therefore, was innocently condemned—a victim of his accuser's and his judge's limited knowledge.

Chapter Second.

MAGNETIC EXPERIMENTS—PHYSICAL PHENOMENA—SLATE-WRITING UNDER
TEST CONDITIONS.

The facts testified to by Mr. Wallace and other well-known Englishmen, as observed by them in the presence of Slade, I can fully confirm on the ground of an investigation of more than eight days with the latter in my own house. As witnesses of the phenomena then observed, and about to be particularly described, I am expressly authorised to cite my friends Professor Fechner, Professor Wilhelm Weber, and Professor W. Scheibner.

On the 15th November 1877, at five o'clock in the afternoon, Slade came to Leipzig for the first time, and took a room in the Palmtree Hotel (*Palmbaum*), which had been ordered for him by two of my friends, at whose invitation he had come here from Berlin. Although I was not a stranger to the literature of Spiritualism, I had hitherto declined to occupy myself personally with its asserted phenomena, because, in the first place, I was quite satisfied to leave these for the present in the hands of two such excellent and unprejudiced observers as Crookes and Wallace; and, secondly, because my time was already fully occupied with my physical researches. Still, I had no reason for refusing the request of my friends to use so convenient an opportunity as the present, and at least to have a look at Slade. I therefore accompanied my two friends on a visit to him on the evening of his arrival, without the least intention, however, of taking part in a sitting, or even of arranging one.

Slade came alone to Leipzig. He had left his niece (the daughter of his deceased wife's sister) as well as his secretary, Mr. Simmons, and his daughter—which three persons accompanied him on his travels—in Berlin, at the Hotel Kronprinz; these persons are, therefore, wholly unknown to me.

The personal impression which Slade made upon me was a favourable one. His demeanour was modest and reserved, and his conversation (he spoke only English) was quiet and discreet. The conversation soon turned upon Lankester's accusation, and his manner and language indicated moral indignation at the proceedings against him in England. To change the subject, I asked him whether he had ever tried to influence a magnetic needle, for I remembered that Professor Fechner had observed a similar phenomenon with Erdmann, late professor of chemistry at the Leipzig University, in the presence of a certain Madame Ruf, a sensitive whom Reichenbach had introduced to those gentlemen.

To give my readers here the interesting result of that investigation, the following is quoted from Fechner's small pamphlet, "Recollections of the Last Days of the Science of Od and its Authors," which appeared two years ago (Leipzig, Breitkopf & Härtel, 1876), under the heading, "Experiments with Madame Ruf."

Fechner's Magnetic Experiments with a Sensitive.

"Saturday, 4th July 1867.— Early to-day Herr von Breichenbach surprised me with a visit. To my repeated refusal by letter to join in his experiments, after I had been unable to obtain a commission from my colleagues to examine the same, and the experiment with the pendulum had come to nothing, he had replied that he would come notwithstanding, and would even bring with him a sensitive to show me the experiments, without claiming from me a public judgment upon them, naturally presuming that I would not avoid giving it if called upon by him for it, supposing only that I had first convinced myself.

"I received him very coldly, explained to him again that I had desired to abstain from a participation in his experiments, of which nothing would come even for himself: but as he was there, I went with him to his hotel, where he introduced to me his sensitive, a large but rather lean woman, between forty and fifty years old, who might once have been handsome; and

I saw a table on which he had laid out all possible preparations—magnets, sulphur and metals melted down in pipes, a raw and a boiled egg, and so forth, as far as I knew.

"The sensitive explained that she was not quite well, and her sensibility was not very highly developed this day.

"An experiment which Reichenbach himself conducted, while I was with him at the hotel, surprised me, and I did not know what to make of it. A common box-compass, with a needle some inches long, under glass, was placed on the table. He caused the sensitive to move a finger to and fro before one of the poles (not over the glass, but in front of the case), and thereby the needle began to oscillate, as if an iron or magnetic rod had been similarly passed before the same pole. These oscillations were not inconsiderable, and the experiment succeeded with each repetition, even when Reichenbach was in other parts of the room, and also when the finger alternately approached and removed from the pole. Trying the experiment in like manner, myself, the needle remained quite motionless. Reichenbach said the phenomenon was weak that day; at times the sensitive had drawn the magnetic needle completely round. I examined the finger in its extent and under the nails as closely as possible, caused the arm to be bared to above the elbow, in order to discover any iron or puncture through which needles could be passed under the skin; in vain. However, I reserved myself for re-examination of this experiment.

"July 13th.—Since the last experiments, the sensitive had fallen into such a condition of insensibility, that Reichenbach, as he wrote me, could stick needles down to the blood in her limbs without her feeling anything. Early to-day he came to me, and said his sensitive was not sufficiently recovered for a repetition of the experiments with the horse-shoe, barmagnet, or pendulum; but the deviation of the magnetic-needle, which had ceased during her state of insensibility, again succeeded; and he begged me to satisfy myself of it immediately, as he was not sure of the continuance of the existing conditions. So I went with him. The magnetic experiments, to which I confined myself, were so successful, that my understanding, so to speak, was in suspense, notwithstanding I endeavoured to exclude all possible means of deception.

"In the foregoing experiments, the sensitive had sat in front of the magnetic needle; this time I made her sit at the side. If the sensitive had had a magnet under her clothes, a suspicion which could be entertained and was all the more to be reckoned with, as it had been very seriously suggested from highly respectable quarters, this would have established quite different conditions of motion of the needle from the former, and rendered generally impossible the regular phenomena which I observed; and even without the pointing of the finger must of itself have produced irregularities in the motion of the needle,—nothing of which happened. Such a suspicion after this could not be maintained. I throughout examined whether the motion of the magnetic needle indicated attraction or repulsion, and it generally appeared that whatever part of the left or right hand or of the arm was applied, the south pole of the needle was repelled, the north pole was attracted; notwithstanding Reichenbach, who appears to have instituted the experiment with the magnetic influence quite superficially, to my question, whether the polar characteristics were distributed to the right and left respectively, so that the one attracted what the other repelled, equilibrium resulting from their joint action, had replied that would indisputably so appear; whereas, in fact, right and left were quite alike in this respect, only the left seemed to act more strongly than the right. A proof, at any rate, that Reichenbach himself had nothing to do with any trick; the phenomenon contradicting his theory, and he being unable to give any definite explanation of it. Reichenbach stood throughout so quietly and at such a distance that there was nothing to guard against from him; and as to the sensitive, I never remarked a motion of the body to support the suspicion that she had a magnet under her clothes, by the motion of which the results were brought about. Moreover, I frequently made her try the experiment only with the finger, expressly bidding her keep her whole body at the same time as quiet as possible; nor could I, with the closest attention, perceive that she disobeyed me.

"After all, one cannot suppose that the woman had stuck needles under the skin in all her fingers, and up to the elbow; and, moreover, only magnetic needles, and these everywhere with a like direction of the pole. Again, as to the suspicion that she contrived the magnetic phenomena of the needle by the motion of a magnet under the clothes, that is entirely excluded, for the reason that the increase or disturbance of the oscillations of the needle, according to the approximation or removal of the finger (with the principle of which act the sensitive was unacquainted, Reichenbach himself not knowing the right application), were exactly such as they must have been supposing a magnetic property in the finger; a result which could not have been produced by art, even if the sensitive had known the principle.

"July 14th.—At eleven o'clock this morning I repeated the experiment with the magnetic needle, in company with Professor Erdmann, whom I had in the meanwhile been able to induce to take part in it. It resulted as before, and Professor Erdmann was impressed as well as myself. Any means of deception could be discovered to-day as little as on former occasions. I had before asked the sensitive whether she had not iron about her, and she had said she had not, but neither she nor I had then thought of her crinoline; to-day, however, she mentioned of her own accord that the experiment succeeded just as well without the crinoline as with it, and offered, as she then had it on, to take it off, which she did in the room. And in fact, the experiment was as completely successful as before. Moreover, it will at once be seen that the earlier-described results, even if possibly influenced by the crinoline, would much rather have been disturbed in their regularity than induced by it. In addition, Reichenbach declared that he was ready to let the experiment be undertaken by ladies whom we might appoint, the sensitive being wholly divested of her clothes.

"*Postscript.*—The following day the woman became so ill that Reichenbach was obliged to send her back; and even later on, she was not fit for the experiments. At her second visit here I recommended her, if the magnetic power returned to her, to introduce herself to some physicist or physiologist by profession, for the purpose of experimentation, and she might thus become a celebrated person; but I have heard nothing more of her.

"Yet are the magnetic results obtained with Madame

Ruf generally so novel (*unerwartet*), that with regard to the hitherto proved impossibility of reproducing them with others, every doubt of their genuine character must be permitted. Was there actually no deception in them? That Reichenbach himself was incapable of wilful deception every one will admit, who from personal intercourse with him, or from reading his writings, knows that he was much too possessed with the reality of the facts adduced by him to hold it necessary to resort to any artifices in support of their credibility; and that even the sensitive herself was not intentionally deceiving, may be inferred from the fact that she throughout presented herself only as a passive instrument of Reichenbach in the experiments, and manifested rather a passive than an active interest in them, as appears from the above accounts. But even should the intention to deceive be presupposed either in him or her, I am absolutely at a loss how such deception could have held out against the altered conditions of the experiments, as I have described them. Could the experiments have been continued, doubtless yet other means of control would have been instituted; *but at least, for my own part, I confess myself convinced already by that which I have been able to communicate hereupon.* It may be thought an hallucination on my part, and indeed I asked myself repeatedly whether I saw rightly; but Professor

Erdmann, whom unfortunately, since his departure, I can no longer call as a witness, must have shared it likewise."

The above fact, established by two well-known reliable witnesses (Professor Fechner and Professor Erdmann), of an influence exercised by a human being upon a magnetic needle, is so remarkable, and stands so wholly outside our ordinary experience, that it must be a matter of the highest interest to every *true* investigator of nature to be able to confirm and repeat this fact with another individual. I therefore put the question to Mr. Slade, whether he experienced anything similar in himself. Slade answered me that last Sunday (11th November 1877) he had been examined as to this peculiarity by a Berlin professor (whose name he did not remember), and on that occasion, the power which he did not know himself to possess, of diverting a magnetic needle and putting it in lively oscillation, had manifested itself. This account first awakened in me the desire to experiment with Mr. Slade in like manner as Fechner had done ten years before with the above-mentioned Madame Ruf.

As I was expecting Fechner and Wilhelm Weber on the following evening (Friday, 16th November) at a small party of friends who assembled every week at my house, I invited Mr. Slade to come and take a cup of tea with us. I explained to him that we should be quite satisfied if he could produce nothing but the divergence of a magnetic needle under conditions which would preclude all notion of suspicion even for the most distant bystanders. Slade accepted my invitation, and was even ready to come *at once* to my house in company with one of my friends. I wished to make sure of the experiment that evening, in order to guarantee its success the following day in the presence of my friends. This intention I, of course, did not intimate to Slade.

Arrived at my dwelling, my friend asked whether I had a compass at hand. I brought a celestial globe in the stand of which a compass was fixed, and placed it on the table. At our request Slade moved his hand horizontally across the closely-fitted glass cover of the magnet case. The needle remained immovable, and I concluded from this that Slade had no magnet concealed beneath his skin. On a second trial, which was made immediately afterwards, in the manner stated, the needle was violently agitated in a way which could only be the result of strong magnetic power.

This observation decided my position towards Mr. Slade. I had here to do with a fact which confirmed the observations of Fechner, and was, therefore, worthy of further investigation.

The next evening (Friday, November 16th, 1877) I placed a card-table, with four chairs, in a room which Slade had not yet entered. After Fechner, Professor Braune, Slade, and myself were seated, and had placed our interlinked hands upon the table, there were raps on the table. Two hours previously I had bought a slate and marked it; on this the writing began in the usual manner. My pocket-knife, which I had lent to Slade to cut off a fragment of pencil, was laid upon the slate, and while Slade was placing the slate partially under the flap of the table, the knife was suddenly projected to the height of one foot, and then thrown down upon the table, but, to our extreme surprise, was open. The experiment was several times repeated with like result, and for proof that the knife was not projected by any movement of the slate, Slade laid at the same time as the knife a bit of slate-pencil on the slate, and, to fix its position, made a small cross on the place. Immediately after the knife had been projected, Slade showed us the slate, on which the bit of pencil remained unmoved near the mark.

The double slate, after being well cleaned and a piece of pencil placed in it, was then held by Slade over the head of Professor Braune. The scratching was soon heard, and when the slate was opened, a long piece of writing was found upon it.

While this was going on, a bed which stood in the room behind a screen suddenly moved about two feet from the wall, pushing the screen outwards. Slade was more than four feet distant from the bed, had his back turned towards it, and his legs crossed, always visible, and towards the side away from the bed. I then returned the bed to its original place.

A second sitting took place immediately with Professor Weber, Scheibner, and myself. While experiments similar to those first described were being successfully made, a violent crack was suddenly heard, as in the discharging of a large battery of Leyden jars. On turning, with some alarm, in the direction of the sound, the before-mentioned screen fell apart in two pieces. The strong wooden screws, half an inch thick, were torn from above and below, without any visible contact of Slade with the screen. The parts broken were at least five feet removed from Slade, who had his back to the screen; but even if he had intended to tear it down by a cleverly-devised sideward motion, it would have been necessary to fasten it on the opposite side. As it was, the screen stood quite unattached, and the grain of the wood being parallel to the axis of the cylindrical wooden fastenings, the wrenching asunder could only be accomplished by a force acting longitudinally to the part in question. We were all astonished at this unexpected and violent manifestation of mechanical force, and asked Slade what it all meant; but he only shrugged his shoulders, saying that such phenomena occasionally, though somewhat rarely, occurred in his presence. As he spoke, he placed, while still standing, a piece of slate-pencil on the polished surface of the table, laid over it a slate, purchased and just cleaned by myself, and pressed the five spread fingers of his right hand on the upper surface of the slate, while his left hand rested on the centre of the table. Writing began on the inner surface of the slate, and when Slade turned it up, the following sentence was written in English : " It was not our intention to do harm ; forgive what has happened." We were the more surprised at the production of the writing under these circumstances, for we particularly observed that both Slade's hands remained quite motionless while the writing was going on.

The above-mentioned phenomena, which we witnessed at our first meeting with Slade, appeared to me and my friends so extraordinary, and so much at variance with all our former conceptions, that William Weber and myself resolved to give some of our colleagues the opportunity of testifying to them. We therefore went the next day to Professor C. Ludwig and informed him of the facts. The interest which he took in the subject encouraged me to invite two other friends to come to my house the next day (Sunday, November 18th), to judge for themselves in the presence of Slade. I proposed my colleagues, Herr Geheimrath Thiersch, surgeon, and Herr Wundt, Professor of Philosophy, in which choice Herr Ludwig fully concurred.

On Sunday, the 18th November, at three o'clock in the afternoon, these three gentlemen met at my house. I had purchased the previous day a *new* walnut-wood card-table from a cabinetmaker in this town, named J. G. Ritter, and had put it in the place of the table used at the former sitting. The slates, single and folding, which we placed at Slade's disposal were bought by myself and my friends, and were marked by us. There were present at the *séance* only Herr Geheimrath Thiersch, C. Ludwig, and Professor Wundt: after half an hour's sitting, they left the room. Of the phenomena observed by them I will only mention that related to me by Herr Thiersch, viz., a successful experiment similar to my own with my pocket-knife, and, in addition, that between the folds of a double slate, which Slade held in his right hand *over* the table in view of all, three sentences were written in the English, French, and German languages, each one in an entirely different handwriting. The slate remains in my possession, and affords opportunity for investigation with regard to the question of previous preparation.

It is to be understood that the present relation of facts in no way presupposes a judgment in the minds of my colleagues as regards the causes of the phenomena. I perfectly agree with the Imperial Court conjuror, Herr Bellachini, whose testimony concerning Slade begins with the following words:—

"I hereby declare it to be a rash act to form any conclusion with regard to the objective mediumistic performances of the American, Mr. Henry Slade, even with the minutest observation, after one sitting only." [See Appendix B.]

Slade returned the same afternoon, about six o'clock, to Berlin. All that had been observed in his presence appeared to me and my friends to be of so interesting a nature, and so entirely worthy of further investigation, that we thankfully and willingly accepted the offer of my friend, Mr. Oscar von Hoffmann, to invite Slade to spend a longer time in Leipsic as his guest, that he might be thus withdrawn from all publicity, and placed entirely at our disposal for the purposes of scientific research. In consequence of this invitation, Slade came a second time alone to Leipsic, on Monday, 10th December 1877, and took up his appointed quarters in the house of my friend.

Next morning (Tuesday, 11th December) at half-past eleven Slade came to my house. This was high and detached, and I had placed the above-mentioned card-table in a corner-room which had four large windows, three to the south, and one to the west. Professor W. Weber, Professor Scheibner, Slade, and I, seated ourselves forthwith at the card-table, which was quite detached, and placed in the middle of the room. Weber was opposite to me, Scheibner at the left, Slade at the right. While our eight hands were upon the table, in contact, and Slade's feet, crossed sideways, were continually observed by the sitter at the side next him, a large hand-bell which had been put under the table suddenly began to ring, and was then violently projected before all our eyes about ten feet distance horizontally upon the floor. After a short pause, in which phenomena similar to those already described took place, a small note-table, fixed to a door-post by a movable iron support, began suddenly to move, and so violently, that a chair standing in front of it was thrown down with a great noise. These objects were behind Slade, and at least five feet from him. At the same time, and at the like distance, a book-case, loaded with many books, was violently agitated. A small paper thermometer-case was laid on the slate, which Slade held half under the edge of the table. This disappeared, so that Slade could show the slate empty; after about three minutes it came again into view upon the slate. Both here, and in the following account, I take no notice of the continually repeated writing between the slates.

On the same day, the same persons assembled in the same room for a second sitting. W. Weber placed on the table a compass, enclosed in glass, the needle of which we could all observe very distinctly by the bright candlelight, while we had our hands joined with those of Slade (which were *both* visible, and over a foot distant from the compass). After about five minutes the needle began to swing violently in arcs of from 40° to 60°, till at length it several times turned completely round. Slade now got up, and went from the table to the window; he hoped that the movements of the needle (which were especially remarkable by reason of the frequent sudden revolutions and the resting points) would be continued in his absence: this, however, did not happen. But when, standing, he again put his right hand with ours (always joined to his) in motion (Slade's hand, however, remaining at least a foot and a half from the compass), the peculiar agitations of the needle suddenly recommenced, and were finally changed into rotations.

In order to repeat some observations with an accordion, in the presence of Home (which were made and published by Crookes and Huggins), besides the above-mentioned large hand-bell, an accordion had been brought by one of my friends. The bell was placed under the table, as in the morning, and Slade grasped the *keyless* end of the accordion (which he had never had in his hands before, but saw now for the first time) above, so that the side with keys hung down free. While Slade's left hand lay on the table, and his right, holding the upper part of the accordion *above* the table, was visible to us all, the accordion began suddenly to play, and at the same time the bell on the floor to ring violently. The latter could thus not be touching the floor with its edges during the ringing. Hereupon Slade gave the accordion to Professor Scheibner, and requested him to hold it in the manner above described, as it might possibly happen that the accordion would play in his hand also, without Slade touching it at all. Scarcely had Scheibner the accordion in his hand, than it began to play a tune exactly in the same way, while the bell under the table again rang violently. Slade's hands meanwhile rested quietly on the table, and his feet, turned sideways, could be continually observed during this proceeding.

Encouraged by the success of this exactly-described experiment, Slade renewed the repeated attempt, hitherto in vain, to obtain writing on a slate held by another, and not touched at all by himself. He therefore handed to Professor Scheibner one of the slates purchased by myself and kept in readiness, requesting him to hold it at first with his left hand under the table, while Slade held it firmly at the edge with his right hand: Scheibner could thus always judge from a pull or pressure whether Slade was holding the slate close under the table. Scheibner's right hand and Slade's left rested meanwhile on the table. After waiting vainly for a short time, Slade remarked that he felt a damp body touching the hand that held the slate, and at the same time Professor Scheibner also testified to the same sensation, which he likened to the touch of a piece of damp felt-cloth. Scheibner then withdrew the slate, which in fact was freely moistened on the *upper side*, both in the centre and at the edges for a breadth of from two to three inches, as were also the hands, both of Scheibner and Slade, which had held the slate.

While we were conjecturing in what conceivable manner this moistening could have happened, and all our hands were on the table, there appeared suddenly a small reddish-brown hand at the edge of the table, close in front of W. Weber, and visible to us all, which moved itself vivaciously and disappeared after two seconds. This phenomenon was several times repeated.

In order, conclusively, to establish the elevation above the floor of one body sounding against another, I had suspended a steel-ball, of about three-quarters of an inch diameter, by a silk thread inside a cylindrical glass-bell of one foot in height and one-half foot diameter. The bell so formed was placed under the table instead of the other bell, and very soon there began a lively tinkling with unmuffled tones as the steel-ball struck against the glass side. As Slade's hands were on the table, and his feet were observed; and even in case of an application by the latter, the tone of the bell would have been affected by the contact of another body; this phenomenon could only be brought about by an elevation of the bell to freedom from contact.

On the next day, the 13th December 1877, Slade proposed to us himself that we should make a direct observation of the movement of the said bell under the table, and thereby make sure that this movement happened without any contact on his part. With this view we sat at a distance of about four feet from the table; by means of candles suitably placed we could conveniently observe everything which happened under the table. The glass-bell was now placed under the table, and indeed towards the side facing us, about in the line between the two feet of the table which were nearest us. Slade sat on the opposite side, and had his feet, visible to us all, drawn back under his chair, so that they were about three feet from the bell. After a short time the bell, without any touching on Slade's part, began moving violently, rolling about in an oblique position upon the lower glass edge, the steel ball thereby grinding against the glass side.

On this evening occurred writing between a double slate, bound cross-wise by a tight knot, and laid on a corner of the table, and which *no one touched.* This result may be compared with that obtained at St. Petersburg, recorded in an English journal, *The*

Spiritualist of March 1st, 1878, which contains the following paragraphs under the title, "Dr. Slade's *Séances* with the Grand Duke Constantine :"—

"On Wednesday last week, Dr. Slade, accompanied by M. Alexandre Aksakow and Professor Boutlerof, gave a *séance* to the Grand Duke Constantine. The Duke gave them a cordial reception, and after a few minutes' conversation, the manifestations began with great power. The Duke held a new slate, alone, and obtained independent writing upon it.

"The Grand Duke Constantine has before this shown his appreciation of new branches of science. When Lieutenant Maury was obliged to flee from the United States during the late civil war, the Duke recognised the—then scarcely appreciated—value of his researches on the physical geography of the sea and oceanic currents, so offered him a home and a welcome in Russia.

" Dr. Slade is fully engaged in St. Petersburg, and sometimes obtains messages in the Russian language. At one of his sittings last week he obtained writing in six languages upon a single slate."

The above fact is additionally confirmed by the following public testimony by M. Aksakow, Imperial Privy Councillor :—

"I can, as a witness, testify that the writing was produced upon a slate which the Grand Duke *alone* held under and close to the table, while Slade's hands were on the table and did not touch the slate. Slade has since had the honour of being invited to two *séances* by the Grand Duke.—AKSAKOW."

The above experiment, described as succeeding with the Grand Duke Constantine, was never successful in my sittings, although Mr. Slade with this object has repeatedly given the slate to be held alone by Professor W. Weber and Professor Scheibner. On the other hand, that of the evening in question (13th December 1877) which succeeded with W. Weber and me was yet more remarkable. Two slates were bought by myself, marked, and carefully cleaned. They were then—a splinter of about three millimètres thickness from a new slate-pencil having first been put between them—bound tightly together, cross-wise, with a string four millimètres thick. They were laid on, and close, to the corner of a card-table of walnut wood which I had shortly before purchased myself. While, now, W. Weber, Slade, and I sat at the table, and were busied with magnetic experiments, during which our six hands lay on the table, those of Slade being two feet from the slate, very loud writing began suddenly between the *untouched* slates. When we separated them, there was upon one of them the following words, in nine lines, "We feel to bless all those that try (?) to investigate a subject so unpopular as the subject of Spiritualism is at the present. But it will not always be so unpopular; it will take its place among the . . . (?) of all classes and kinds." The slate had the mark (H.2) *previously* placed by me upon it. There can be no talk here of a trick or of antecedent preparations.

In addition, the large hand-bell which was laid on the floor at the side of the table opposite to me, was placed quietly and slowly in my left hand, which I held close under the table; during this proceeding also, Slade's hands were both visible, and his feet were under our control. Finally, Mr. Slade himself proposed an experiment which should serve as proof that the slates were not previously prepared and the writing already present on them, invisibly, before the apparent production of it. He took as usual the slate which came to hand, laid a bit of slate-pencil of the size of a pea upon it, and asked me, while he pushed the slate *half* under the edge of the table (so that his hand could be continually observed), what should be written upon it. I said " Littrow, Astronomer." The usual scribbling began immediately, and when Slade drew out the slate, the two above words were perfectly distinct upon it, with the letters widely apart. If Slade did not write the words himself (at the time), which from the position of his hand and of the letters upon the slate was *impossible*, so likewise could these words certainly not have been produced by means of a previous preparation of the slate, since the words themselves had occurred to me quite suddenly for the first time.

Friday, 14th December 1877 (11.10 to 11.40 A.M.). To-day, first one of the slates kept always in readiness, which I myself selected and cleaned, was laid *open* with a bit of slate-pencil upon the floor under the table. Now, while Slade had both his hands linked with ours upon the table, and his legs, turned sideways, were continually visible, writing, loudly perceptible by us all, began on the slate lying below. When we raised it, there were on it the words—"Truth will overcome all error!" Next, two magnetic needles, a larger and a smaller one, both completely enclosed in glass cases, were placed close in front of W. Weber. Our hands were linked upon the table with those of Slade in the usual manner, and were at least one foot from the magnetic needle. Suddenly the small needle began to oscillate violently, till it got into constant rotation, while the larger one showed only slight agitations, which appeared to proceed from a shaking of the table. Since here, forces were manifestly at work (no matter what their origin) which were able to act upon the magnetism of bodies, I suggested to Slade the attempt, permanently to magnetise an unmagnetic steel-needle. Slade hesitated at first, and seemed to think our success doubtful. However, he was at once ready to consent to the proposition. I fetched a large number of steel knitting-needles, and

W. Weber and I chose from them one which, *immediately* before the experiment (on the table at which we sat), was ascertained by means of the compass to be wholly unmagnetised, inasmuch as both poles were attracted. Slade laid this needle upon a slate, held the latter under the table just in the same way as for writing, and after about four minutes, when the slate with the knitting-needle was laid again upon the table, the needle was so strongly magnetised at one end (and *only* at one end) that iron shavings and sewing-needles stuck to this end; the needle of the compass could be easily drawn round in a circle. The originated pole was a south pole, inasmuch as the north pole of the (compass) needle was attracted, the south pole repelled. The needle is still in my possession, and can at any time be tested.

Chapter Third.

PERMANENT IMPRESSIONS OBTAINED OF HANDS AND FEET—PROPOSED CHEMICAL EXPERIMENT—SLADE'S ABNORMAL VISION—IMPRESSIONS IN A CLOSED SPACE—ENCLOSED SPACE OF THREE DIMENSIONS OPEN TO FOUR-DIMENSIONAL BEINGS.

As almost regularly at all the sittings (while Slade's hands rested on the table, visible to all present, and his feet, in the sideways position frequently mentioned, could be at any time observed) we felt the touch of hands under the table, and, as above remarked, had even seen these transiently under the same conditions, I desired to institute an experiment by which a convincing proof of the existence of these hands could be afforded. I therefore proposed to Mr. Slade to have placed under the table a flat porcelain vase filled up to the edge with wheat flour, and that he should then request his "spirits" to put their hands in the flour before touching us. In this manner the visible traces of the touching must be shown on our clothes after the contact, and at the same time Slade's hands and feet could be examined for remains of flour adhering to them. Slade declared himself ready at once for the proposed test. I fetched a large porcelain bowl of about one foot diameter and two inches deep, filled it evenly to the brim with flour, and placed it under the table. We did not trouble ourselves at first about the eventual success of this experiment, but continued for over five minutes the magnetic experiments, Slade's hands being all the time visible upon the table; when suddenly I felt my right knee powerfully grasped and pressed by a large hand under the table for about a second, and at the same moment, as I mentioned this to the others and was about to get up, the bowl of meal was pushed forward from its place under the table about four feet on the floor. Upon my trousers I had the impression in meal of a large strong hand, and on the meal-surface of the bowl were indented the thumb and four fingers with all the niceties of structure and folds of the skin impressed. An immediate examination of Slade's hands and feet showed not the slightest traces of flour, and the comparison of his own hand with the impression on the meal proved the latter to be considerably larger. The impression is still in my possession, although through frequent shakings the delicacy of the lines is becoming gradually obliterated by the falling together of the particles of meal.

Slade was highly pleased at the success of the magnetic experiments, particularly the magnetising of the knitting-needle, an attempt which we often repeated on the following day with always the like result. He expressed in warm terms his happiness, that he had, for the first time, succeeded in interesting men of sincere inclination to truth for his peculiar endowments, in such a degree that they had resolved to institute scientific experiments with him.

I was now sufficiently encouraged, gradually to set on foot those experiments which I had prepared from the stand-point of my theory of a space of four dimensions. Since the magnetic experiments had proved that under the influences which invisibly surrounded Slade, the molecular currents, present in the interior of all bodies, could be turned, that is, altered in their position (whereon, according to Ampère's and Weber's theory, the magnetising of bodies principally depends), I entertained the hope that an experiment indicated in the first volume of my *Scientific Treatises* would succeed; viz., the conversion, by a four-dimensional diversion of molecules of tartaric acid, which diverts the plane of polarised light to the *right*, into racemic acid, which diverts it to the *left*. To this end I had kept in readiness one of Mitchell's simple polarising saccharometers, the tube of which contained a concentrated solution of tartaric acid. The diversion of the plane of polarisation amounted to about 5°. I intended that the glass tube (200 millimètres long and 15 outer diameter), filled with the solution, should be laid on the slate, the latter being held by Slade under the table, as in the case of the knitting-needles which were to be magnetised; in the expectation that after the experiment I should see the tartaric acid changed into racemic acid. Wishing first to explain to Mr. Slade the meaning of the experiment, I began by pointing out to him in the apparatus itself, after removing the tube, the optical effect of two crossed Nicol's prisms. I desired him, while sitting in his chair, to fix his eye on the front prism, and then to look with the apparatus at the clear sky (the experiment took place at my house at 11.45 in the morning of the 14th December 1877), while I slowly turned the front Nicol. I now asked Slade, when the two prisms were about crossed, if he observed the gradual darkening of the field of view. To my great surprise, he said he did not. I supposed him to be deceived by the side light, and therefore disposed the two prisms from the front at right angles, so that neither I nor my friends could see through at all. Slade still asserted that he did not perceive the least change in the clearness of the sky; and as proof he read an English writing, placed before the two crossed Nicol's, covering his left eye, as we saw, with his left hand. I was not, however, contented with this proof of the fact. Next morning, when we were again assembled at my house, I had two very large Nicol's prisms (for the production of a greater field of view) fixed to turn closely one over the other, and a large circular screen, which completely covered the sight of the observer, so placed in connection with the prisms, that external objects could only be perceived through the two Nicol's prisms. I then took an English book, Tyndall's *Faraday as a Discoverer*, and in Slade's absence marked by interlineations the following words on page 81 :—"The burst of power which had filled the four preceding years with an amount of experimental work unparalleled in the history of Science." When I again made Slade look through the two crossed Nicol's at the sky, and he declared, as on the day before, that he did not remark the least change in the clearness of the sky when the prisms were turned, I requested him to sit on a chair, and to read to me the underlined words from the book, held at a distance of about two feet from his sight. To the great astonishment of us all, he immediately read the above words with perfect accuracy. When, about ten minutes later, I held the two prisms crossed again before Slade's eye, he was no longer able to see, and the experiment was not more successful in the evening of the same day by candlelight. He informed me that in the morning, soon after the experiment in question, he had perceived "an influence," to which he ascribed the change of his condition. In connection with what has been quoted above, from Professor Fechner, with reference to the change in the magnetic condition of a sensitive, this alteration in Slade's optical powers may afford an interesting confirmation of the transitory character of such anomalous organic functions. The originally intended experiment with the tartaric acid was discontinued in consequence of the above extraordinary observations. I purposed to carry it out at a future investigation of Slade's peculiarities.

On Saturday, the 15th December 1877, at eleven in the morning we assembled again at my house. While we were taking a small breakfast, standing in my work-room, and I was talking to Slade near my bookcase, some twenty feet from the stove, about the experiment with the crossed Nicol's prisms (which Slade designated a "clairvoyant experiment"), there fell suddenly from the ceiling of the room a piece of coal the size of a fist. A similar incident happened half-an-hour later, when my colleague Scheibner, in conversation with Slade, was on the point of leaving the sitting-room; a piece of wood, instead of coal, falling suddenly from the ceiling. On the morning of the 11th December when we stood talking, after the sitting, and I was standing near Slade, we suddenly saw my pocket-knife, fortunately shut, fly through the air, and strike the forehead of my friend Scheibner with some force, the scar remaining visible on the following day. Since at the time of the incident I was conversing with Slade, and the latter had his back turned to my friend at a distance of about ten feet, Mr. Slade, at any rate, could not have thrown the knife at my friend's head. I only cite this incident because it appears to me to belong to the same class as the above-mentioned facts.

Those experiments seem to me far more important, however, in which *permanent* impressions of contact were left behind, as was the case with the impression of the hand in the bowl of flour.

With this view I stuck half a sheet of common letter-paper upon a somewhat larger board of wood; it was the cover of a wooden box, in which Herr Merz had sent me some large prisms for spectroscopic purposes from Munich four days before. By moving the paper over a petroleum lamp without a cylinder it was spread all over with soot (lamp black), and then placed under the table at which W. Weber, Slade, and I had taken out seats. Hoping to obtain upon the sooted paper the impress of the hand, as on the previous day, we at first directed our attention again to the magnetic experiments. Suddenly the board was pushed forward with force under the table about the distance of one meter, and on my raising it, there was on it the impression of a *naked* left foot. I at once desired Slade to stand up and show me both his feet. He did this most willingly; after he had drawn off his shoes, we examined the stockings for any adhering particles of soot, but without finding anything of the sort. Then we made him put his foot on a measure, from which it appeared that the length of his foot from the heel to the great toe was 22·5 centimètres, whereas the length of the impression of the foot between the same parts amounted only to 18·5 centimètres.

Two days later, on the 17th December 1877, at eight o'clock in the evening, I repeated this experiment, only with the difference that instead of a board 46 centimètres long by 22 broad, a slate was used, whose surface, not covered by the wooden frame, was 14·5 centimètres broad and 22 long. Upon this free surface I stuck a half sheet of letter-paper (Bath) cut down to exactly the same dimensions. Immediately before the sitting, I myself, in the presence of witnesses, sooted the paper in the manner above described. The slate was then, as before the board, laid under the table at which we sat, with the sooted side uppermost. Upon a given sign we got up after about four minutes, and upon the slate was again the impression of the same left foot which we had obtained two days earlier upon the board. I have had this impression reproduced photographically on a reduced scale.

I learned subsequently, from my colleague Councillor Thiersch, that the method of taking impressions of human limbs on sooted paper was already frequently applied for anatomical and surgical purposes. In the judgment of Herr Thiersch, who had taken a great number of such impressions of feet of different persons for comparison with that obtained by us, the impression produced in the presence of Mr. Slade is that of a man's foot which had been tightly compressed by the make of the shoe, so that, as often happens, one toe is pressed over the two next, and thus only four toes touch the sooted surface on imposition of the foot, as is also the case on the photograph. Herr Thiersch showed me the impression of a human foot in which likewise only four toes appeared in the way denoted. To fix these soot-impressions, it is only requisite to pass them through a thin alcoholic solution of shell-lac. With reference to the greatly abbreviated length of the foot in proportion to its breadth, Herr Thiersch remarked that this could be effected by not putting down the heel and the fore part of the foot at the same time. In fact, he showed me an impression of a foot in which a nearly similar abbreviation had been produced in this way. If upon these observations it should be supposed that Mr. Slade had himself produced the impression by putting on his foot in this way, it must first be assumed that he was able to draw off and on his shoes and stockings without application of his hands (which were all along observed by us upon the table); and secondly, that he was so expert in the imposition of his foot on a narrowly limited space (the surface of the slate), that, *without seeing this surface*, he could, nevertheless, always hit upon it with accuracy. This, certainly, would presuppose a large practice in Mr. Slade for the object intended, and thereby it must be conjectured that he had been used to bring forward this experiment. Putting aside his lively astonishment and his assurance that such phenomena had never yet been observed in his presence,* up to the present time I am not aware of any published accounts of Mr. Slade's production of similar facts.† That Slade's stockings had not been cut away underneath for this purpose —as was conjectured by some "men of science" in Leipzig, who in unimportant things accept our physical observations with absolute confidence, but in reference to the foregoing have not hesitated to instruct us in the elementary rules for instituting exact observations—of that, as already mentioned, we satisfied ourselves immediately after the experiment.

Meanwhile, to meet all such doubts (and the attempts at explanation are scarcely less wonderful than are the facts themselves), I proposed to Mr. Slade an experiment which, according to the theory of the four-

* With reference to this statement, the translator may observe that he has himself had many sittings with Slade, previous to that time, has received accounts of the phenomena occurring in his presence from many who have had equal or greater experience of them, and has read many records of them; yet the above, and nearly all other of the special experiments described in the text (Professor Zöllner's), are wholly new to him.
† To appreciate the importance of this, with reference to the suggestion that Slade is an expert, it is necessary to bear in mind that he has been for many years following his vocation as a medium *in the light of the utmost publicity;* the Spiritualist journals (which are numerous) of America and England having printed innumerable accounts of his *séances.*—TR.

dimensional space, must easily succeed. In fact, if the effects observed by us proceed from intelligent beings occupying (*welche sich befinden*), in the *absolute* space, places which in the direction of the fourth dimension lie *near* the places occupied by Mr. Slade and us in the three-dimensional space,* and therefore necessarily invisible to us, for these beings the interior of a figure of three-dimensional space, enclosed on all sides, is just as easily accessible as is to us, three-dimensional beings, the interior of a surface enclosed on all sides by a line—a two-dimensional figure. A two-dimensional being can represent to itself a straight line with only one perpendicular (*Normale*) in the respective two-dimensional regions of space (to which it belongs phenomenally). We, on the contrary, as three-dimensional beings,

* The conception of the juxtaposition of different, infinitely extended regions of space (*Raumgebiete*) necessarily presupposes the conception of the next higher region of space. Thus a two-dimensional being could indeed conceive any number of parallel infinite straight lines; that is, infinitely extended spaces *(Raumgebiete)* of *one* dimension, but the infinite plane in which it moves, as we with our bodies in the infinitely extended *three*-dimensional space, could represent to itself only *once*, although we, as three-dimensional beings, know that there can be any number of infinitely extended parallel planes, which according to a perpendicular direction, that is, according to the third dimension, can be arranged in *juxtaposition*. All these planes would represent infinitely extended two-dimensional worlds, whose occurrences in each region of space are completely separated from those in another. If, however, under certain anomalous conditions, a two-dimensional being of the one plane were causally connected with more two-dimensional beings of another plane, so that these beings by movements according to the third dimension could produce effects in the two-dimensional region of the first plane, this would seem just as wonderful to the moving beings in the latter as do to us the effects witnessed in the neighbourhood of Mr Slade.

know that there are infinitely many perpendiculars (*Normale*) to a straight line in space, which collectively form the two-dimensional geometrical place of the perpendicular plane of that straight line. Analogously, we can conceive only *one* perpendicular to a plane; a being of four dimensions would, however, be able to conceive infinitely many perpendiculars to a plane, collectively forming the three-dimensional place which in the fourth dimension stands perpendicular to that plane. By our nature as three-dimensional beings we could form for ourselves no *representation* of these space relations, although we are in the position to discover ideally (*begrifflich*), by analogy, the *possibility* of their real existence. The *reality* of their existence can only be disclosed through *facts of observation*.

In order to obtain such an *observed fact*, I took a book-slate, bought by myself; that is, two slates connected at one side by cross hinges, like a book for folding up. In the absence of Slade I lined both slates within, on the sides applied to one another, with a half sheet of my letter-paper which, immediately before the sitting, was evenly spread with soot in the way already described. This slate I closed, and remarked to Mr. Slade that if my theory of the existence of intelligent four-dimensional beings in nature was well founded, it must be an easy thing for them

to place on the interior of the closed slates the impressions of feet hitherto only produced on the open slates. Slade laughed, and thought that this would be absolutely impossible; even his "spirits," which he questioned, seemed at first much perplexed with this proposition, but finally answered with the stereotyped caution, "We will try it." To my great surprise, Slade consented to my laying the closed book-slate (which I had never let out of my hands after I had spread the soot) on my lap during the sitting, so that I could continually observe it to the middle.*
We might have sat at the table in the brightly-lighted room for about five minutes, our hands linked with those of Slade in the usual manner *above* the table, when I suddenly felt on two occasions, the one shortly after the other, the slate pressed down upon my lap, without my having perceived anything in the least visible. Three raps on the table announced that all was completed, and when I opened the slate there was within it on the one side the impression of a *right* foot, on the other side that of a *left* foot, and indeed of the same which we had already obtained on the two former evenings.

My readers may judge for themselves how far it is possible for me, after witnessing these facts, to con-

---

* In the previous experiments the board and the slate had been laid open upon the floor under the table.

sider Slade either an impostor or a conjurer. Slade's own astonishment at this last result was even greater than my own. Whatever may be thought of the correctness of my theory with regard to the existence of intelligent beings in four-dimensional space, at all events it cannot be said to be useless as a clue to research in the mazes of Spiritualistic phenomena.

Chapter Fourth.

CONDITIONS OF INVESTIGATION—UNSCIENTIFIC MEN OF SCIENCE—SLADE'S ANSWER TO PROFESSOR BARRETT.

PASSING over the numerous other physical phenomena, such as violent movements of quite unattached chairs and the like, since the same have been so often observed and circumstantially described by others, I may next discuss the question how far it is justifiable and reasonable in dealing with *new* phenomena, the causes of which are entirely unknown to us, to *impose conditions* under which these new phenomena should occur. That for the production of electricity by friction on the surfaces of bodies the driest possible air is requisite, and that in a damp atmosphere these experiments fail entirely, are also experimental conditions, which could evidently not be prescribed *a priori*, but have been discovered only through careful observations among those relations under which Nature in individual cases willingly offers us these phenomena. Just therein, indeed, consists the acuteness and skill of an observer, that without arbitrary meddling with the course of the phenomena, he so prepares his observations that the conclusions drawn from them exclude the possibility of every error and every deception.

Would it have been possible to dictate conditions under which the fall of meteorolites should be observed, upon those who first asserted the reality of those phenomena? On entering new provinces one must always take to heart the words of Virchow, which he uttered at the last meeting of scientific men at Munich, in his speech "Upon the Freedom of Science in the Modern State."

"That which I pride myself on is just the knowledge of my ignorance. Since, as I imagine, I know with tolerable accuracy what it is that I do not know, I always say to myself, when I have to enter upon a province as yet closed to me, '*Now must thou begin again to learn!*'"

How far Herr Virchow himself, when the occasion is forthcoming, makes use of the teachings of modesty which he imparts to others, we may learn from the following words of Herr State Counsellor Aksakow:*—

"The attempts which I caused to be made by Herr Wittig in Berlin for a scientific examination of Mr. Slade by Professors Helmholtz and Virchow have failed; and I will take this opportunity to show by an example how right I was in speaking of the difficulties which we still have to experience with the learned, even when it is a question of simply putting

* *Psychische Studien*, monthly journal devoted principally to the investigation of the little-known phenomena of the soul-life. Published and edited by Alexander Aksakow, Russian Imperial Counsellor of State, at St. Petersburg,—January number, 1878.

the mediumistic phenomena to the proof, and this solely by reason of their disinclination for this province of investigation. Thus, Herr Virchow is willing indeed to see Mr. Slade, but only upon the terms that the latter submits himself to *all* conditions which Herr Virchow shall please to lay upon him. Here now is a man of science (*Gelehrter*) who, not knowing even the A B C of the phenomena which he undertakes to make an object of his study, at the outset imposes upon them his own conditions of observation! Could a similar method have been at all approved or endured in the study of any branch of natural science whatsoever? ... So the *first* false step! And then what were these conditions? Mr. Slade should allow Professor Virchow to bind his hands and feet, and to place an observer two feet from the table. These are the conditions required by a German man of science of great renown, and, nevertheless, how 'illogical and inconclusive' ('*unlogisch und beweis-unkräftig*') are they! Take it that Mr. Slade submits to these conditions, and the *séance* is successful. Herr Virchow will be the first, and with him the whole great multitude, thence to conclude that *he had tied badly*, that his sentinel *had observed badly*, and that the adroitness of the conjuror had taken him at a disadvantage. At a second *séance* Herr Virchow will bind the medium in a different manner, and will appoint two sentinels; the same result, the same conclusion! At the third *séance* he will discover yet another system of fastening and precautions much more elaborate and ingenious; the same result, the same conclusion, and so on for ever!*

* Even if the above supposition is thought unjust to Professor Virchow (as it perhaps is), it is one which Slade's past experience made a reasonable ground for the rejection of the Professor's conditions. When Slade was in London in 1876, a distinguished man of letters was anxious to obtain writing in a new book-slate furnished with a padlock, and locked before it was brought to the *séance*. Slade declined the attempt, greatly to the dissatisfaction of the gentleman referred to, whose distrust on this account was reflected in the tone of his evidence at Bow Street, on the charge against Slade by Professor Lankester, though otherwise he was witness to inexplicable manifestations. On my urging Slade subsequently to comply, he told me that this very test had once been successfully allowed, but that the fact getting known, it had led to other new contrivances being devised and insisted on, with an utter disturbance of the usual conditions. He never could be sure beforehand that a *séance* would succeed (the manifestations being wholly out of his own power or control), and the failure of a test imposed by the investigator was regarded as more suspicious than many merely weak and inconclusive *séances* under ordinary conditions. (See also Slade's letter to the *Times*, page 67 post.) There is also the fact, well recognised amongst Spiritualists, that the influence of some persons is far more favourable to the evolution of phenomena through mediums than that of others. One investigator will witness the most extraordinary manifestations at his first *séance*, whereas another will be long in obtaining anything like satisfactory evidence, as was the case with myself before I saw Slade. This interaction of medium and sitter is a fact that should never be left out of sight; especially in estimating testimony to facts far exceeding our own or general experience of similar phenomena. It by no means followed from Professor Zöllner's success in nearly all the experiments he instituted with Slade, that another man of science, of perhaps altogether different constitution, physical or psychical, would be equally fortunate. The true cause of scientific complaint against Prof. Virchow appears to be that he would not even *in the first instance* witness the phenomena under the ordinary conditions of their occurrence; assuming that there could be only one mode of demonstrating them to be genuine, or that, out of many modes, that which occurred to him must also be agreeable to nature. Probably he only thought how *he* could baffle a conjurer, not entertaining the possibility that the very course and nature of the phenomena themselves might put the hypothesis of conjuring out of the question.—TRANSLATOR.

Mr. Slade did well to decline Herr Virchow's conditions; for in imposing them the latter had given proof of an utter ignorance of the subject which he professed his willingness to engage in. The history of all the systems of fastening by which mediums have been tortured would alone fill a thick volume. The *Martyrology of Mediums* is a book of the future. . . . Professor Virchow need only open the book by Colonel Olcott—*People from the other World*—at page 39, to see a pictorial representation of the tortures to which mediums have been subjected in the name of science and truth. There is represented the medium Eddy, with every finger of the hand separately fastened by a string nailed to the floor. Eddy's hands are in consequence of these bindings, to which they have been subjected for years, quite disfigured. And have all these bindings ever convinced any one? The conditions devised by Professor Virchow would have the same fate.

"Slade's great merit is to have simplified the conditions of his *séances* in such a manner that it is sufficient for any one to come to him armed only with his sound senses and with his sound reason to be convinced—if he *will* be convinced. In fact, the phenomena take place in full light, and while the medium's hands and feet are held,* or, also, when the

---

* That is, when Slade does not himself hold the slate partly under the table. He is always willing to use new slates, brought by the visitor,

medium does not even touch the object upon which the mediumistic phenomena are accomplished, and while the observer does not cease to hold both his hands, and to see with his own two eyes! What more is necessary?"

I cannot refrain from setting down here the letter, full of sound manly sense, which Mr. Slade sent to the *Times* in London, in reply to some points raised by Professor Barrett of Dublin:—

Dr. Slade's Answer to some Points of the Letter of Professor Barrett.

"London, 8 Upper Bedford Place,
"*September 22nd*, 1876.

"Sir,—In Professor Barrett's statements published in the *Times* to-day I think he erred (I hope unintentionally) in saying:—'Slade failed to procure the writing on a slate enclosed, along with a fragment of pencil, in a sealed box; he also failed when I used a box with a tortuous passage to allow the introduction of such bits of pencil as suited his fancy; he declined to try and get writing within a hinged slate that was sealed, but succeeded when the hinged slate was unfastened; and again he failed, according to the

on which writing is often obtained *above* the table. The full light is an invariable condition. The most conclusive tests cannot, however, be insisted on arbitrarily, at once, always, and by any one, but are usually given in the course of a few sittings.—Translator.

writer of an article in the *Spectator*, when a spring lock was used.'

"Instead of trying to obtain writing on the Professor's boxed slates, I declined using them at all. I assured him they would not be used, and gave him my reasons for objecting. He urged me strongly to make the experiment, and placed the box containing the slate on the table, where it remained undisturbed until he put it on the slate, which I held, with the box on it, under the table for a short time, when, as I had hoped, nothing occurred. This he calls a failure.

"Mr. Simmons says that Professor Barrett, on entering the drawing-room after the sitting, told him that Dr. Slade had refused to use the boxed slates; that he had left them in the room where the sitting was held, hoping he (Dr. Slade) would make the trial at some future time.

"Having had at least fifteen years' experience in demonstrating the fact of various phenomena occurring in my presence, I claim to know something of the conditions required. At the same time I do not know how they are produced. I do not object to persons bringing an ordinary slate, either single or folding, but I do object to using locks, boxes, or seals, for this reason—I claim to be as honest and earnest in this matter as those who call upon me for the purpose of investigation. Therefore I shall continue to object to all such worthless appliances whenever they are proposed.

"Mark the following, which Professor Barrett also says :—'Taking a clean slate on both sides, I placed it on the table so that it rested above, though it could not touch, a fragment of slate-pencil. In this position I held the slate firmly down with my elbow. One of Slade's hands was then grasped by mine, and the tips of the fingers of his other hand barely touched the slate. While closely watching both of Slade's hands, which did not move perceptibly, I certainly was much astonished to hear scratching going on, apparently on the under side of the slate, and when the slate was lifted up I found the side facing the table covered with writing. He also says a similar result was obtained on other days; further, an eminent scientific friend obtained writing on a clean slate when it was held entirely in his own hand, both of Slade's being on the table.'

"The above being true, would the fact of the writing being produced by some agency foreign to myself have been more strongly established had it occurred on the Professor's boxed slate? I think the reader will agree with me in saying it would not.

"On the other hand, had it so occurred and a statement of it been published, it would only have served as an incentive for others to conjure up some plan whereby they might prevent an occurrence of phenomena, instead of being content to witness them in the simple manner in which they do occur. To my mind it would be as reasonable to sever the wire and then ask the operator to send your message, as it is to violate the conditions which experience has taught me are essential in these experiments in order to obtain successful results; and when the investigator comes in the spirit of a seeker for truth instead of trying to prove me an impostor, I shall be most happy to unite with him in the further pursuit of these experiments.—Very truly yours,

"Henry Slade."

The above letter, in which the so severely calumniated American medium recalls—in a manner no less urgent than civil—to the recollection of our modern "men of science" the first rules of experimentation in natural science, may suffice for the present to afford the reader an idea of the *intellectual* worth of the man who was sentenced to three months' imprisonment with hard labour, on the charge of fraud brought against him by a young "man of science."

Chapter Fifth.*

PRODUCTION OF KNOTS IN AN ENDLESS STRING—FURTHER EXPERIMENTS—MATERIALISATION OF HANDS—DISAPPEARANCE AND REAPPEARANCE OF SOLID OBJECTS—A TABLE VANISHES, AND AFTERWARDS DESCENDS FROM THE CEILING IN FULL LIGHT.

THE establishment (*Constatirung*) of physical facts falls within the domain of the physicist; and if men of such distinguished eminence as Wilhelm Weber, Fechner, and others, after thorough experimental investigation, publicly attest the reality of such facts, it is evidently nothing but an act of modern presumption for unscientific people, at their pleasure, to accept as facts absurd conjectures concerning the possibility of trickery without more inquiry, and thus to deny the capacity of these men for exact observations.

I have already described in detail the conditions under which the knots [represented in Plate I.] occurred in the string fastened by a seal, in the presence of Mr. Slade, without the string being touched. Every possibility that these knots were in the string already, before the sealing of the ends, and had only been brought to another part of the same by pushing, is hereby definitely excluded.

* *Wiss. Abh.*, Vol. ii., part 2, p. 905.

It will in the first place interest my readers to learn that this experiment succeeded four months later in London in presence of another medium. Under the title, "Remarkable Physical Manifestations," Dr. Nichols has published the following in two letters to the London *Spiritualist* of April 12th and 19th, 1878:—

REMARKABLE PHYSICAL MANIFESTATIONS.

"It may seem tiresome to you to repeat facts, and cumulate evidence, but this appears to be the only way to convince the sceptical. Then you are to consider that each number of the *Spiritualist* falls into the hands of some who have seen no other. So I give you some facts new to me, though they may be familiar to you and most of your readers.

"Busy at my writing the other day in my study, at about two P.M. the housekeeper came with her eyes 'round' with wonder, and begged me to go instantly to the drawing-room over my head. It seemed an urgent case, and I ran upstairs and found every chair but three turned upside down; the large and heavy sofa lying forward in the room, legs upward; and the upright pianoforte prone upon the carpet, flat upon its face.

The windows are sixteen feet from the ground; no person in the house had visited the room that morning; no one could by any possibility have come in from the street to do this work, and it certainly was not done by any of the inmates of the house; at my desk I can hear every footstep in the drawing-room; in a word, it is certain that no visible being had done it. It required two strong men to lift up the pianoforte and restore it to its proper position. The *bouleversement* seems to have been accomplished while most of the family were at lunch, between one and two o'clock; with them were Mr. W. Eglinton and Mr. A. Colman. Mrs. Nichols was with them at table, and reports that, as they were conversing, loud raps responded, and the heavy table, loaded with dishes, when no one touched it, rose up some inches from the floor, and so remained, while she stooped down to see that all its feet were in the air. This is common enough in the presence of mediums, but the very powerful action in the drawing-room, in the light of mid-day, with no person near, seems to me novel and remarkable.

"I gave you some account, I think, of chairs being 'threaded' on the arms of persons while they were firmly holding the hands of others. This is as great a wonder as that reported by the German astronomer at Leipsic—the tying of knots in a cord, the ends of which were sealed together. I have seen the chairs on the arms of seven persons, whose word I could perfectly trust, but I wished to make assurance doubly sure; so at a recent *séance* I tied the two wrists together with cotton-thread. In three seconds the chair was hanging upon the arm of one, and I found the thread unbroken. I then held the hand of Mr. Eglinton as firmly as possible in mine, and in an instant the chair, one of our cane bottoms with bent backs, was hanging on my arm. This, beyond all doubt, was matter passing through matter, but whether the wood passed through flesh and bone, or flesh and bone through wood, I have not yet been able to determine.

"On Saturday, by special appointment, four of us sat at noon—Eglinton, Colman, Mrs. Nichols, and myself. Supposing there might be writing or drawing, I laid a sheet of marked note-paper and pencil on the table around which we sat. It is a small room, and sitting in a good light, we heard a slight noise of something moving, of light raps or knocks in one corner. Looking, we all saw a light cane-bottom chair, about six feet from the table, tilting itself upon two legs, rocking backward and forward, tilting back and balancing on its hinder legs, answering our questions with its movements; and finally, at our request, it walked forward on two of its legs and placed itself at the table, pressed against my knee caressingly, and behaved in all respects like a chair gifted with sense and locomotion. It was a weird spectacle; but it was also a very interesting fact, seen for ten or fifteen minutes by four persons, without the possibility of trick or hallucination. I examined the chair carefully, though it was quite needless to do so, for no conceivable machinery could, under the circumstances, have produced the phenomenon.

"Then the light was turned off for a minute or so, during which we heard rapid movements of a pencil, and on relighting the gas, we found on the marked sheet of paper the portrait of a deceased friend, and a letter of more than a page in the well-known handwriting of a beloved child whose spirit often visits us. I have now from her hand five elaborate drawings and four letters, no one of which occupied two minutes under absolute test conditions. No living artist could make them in from ten to twenty times the time occupied in their production.

"Your readers may be glad to know that, on the night of April 7th, we had repeated, in my house, in the presence of six persons, including Mr. W. Eglinton and Mr. A. Colman, Professor Zöllner's marvel of tying knots in a cord, the ends of which were tied and sealed together. I have the sealed cord, which I prepared myself, with the knotted ends firmly sealed to my card, on which the fingers of every person present rested while five knots were tied, about a foot apart, in the central portion of the cord. I have no doubt that this splendid manifestation can be repeated at any time under like conditions.

"*April 12th*, 1878."

Tying Knots in an Endless Cord.

To the Editor of the "Spiritualist," April 19th, 1878.

"Sir,—I am sorry to learn that my account of the repetition in London of the great Leipsic experiment of tying knots in a cord whose ends were firmly sealed together, was not so accurate as should have been the record of so astounding a phenomenon.

"Permit me to say, therefore, that after reading the account by Professor Zöllner in the *Daily Telegraph*, I asked, at the first opportunity, our spirit-friend, 'Joey,' if he could do the same thing here. He said, 'We will try.'

"I then cut four yards of common brown twine—such as I use for large book packets—from a fresh ball, examined it carefully, tied the two ends together by a single knot, which included both, then passed each end through a hole in my visiting card, tied a square knot, and firmly sealed this knot to the card, and asked a gentleman to seal it with his seal ring. On this card I also put my signature and the date. The loop of the string, whose two ends were thus sealed on the card, I again examined, and found it free from knots.

"Six persons, including Mr. Eglinton and Mr. Colman, sat round a small table. The sealed card was placed on the centre of the table, and the fingers of each person present placed upon it, while the loop hung down upon the floor.

"This position was maintained for about a minute, when raps were heard, and I examined the string. The ends were firmly fastened and sealed as before, and five single knots were tied upon it, about a foot apart—on the single endless string, observe, whose perfect fastening had never left my sight—where they now remain.

"It is certain that no mortal man could have tied these knots—equally certain that all the philosophers and all the 'magicians' of Europe cannot now untie them under the same conditions.

"Here is a fact which can be proven in any court of justice, and for which any conceivable number of dimensions of space cannot acount.

"T. L. NICHOLS, M.D.

"32 FOPSTONE ROAD, LONDON, S.W."

I now pass on to relate, from my numerous successful experiments with Mr. Slade, during his further presence in Leipsic, from 4th to 10th May 1878, those in the first place which represent a modification of the experiments with knots, and which may be regarded as an experimental confirmation of the reality of a fourth dimension of space.

At his third residence in Leipsic, Mr. Slade had again received the hospitable invitation of my friend

Oscar von Hoffmann, and therefore lived in his house during the time from the 2nd to the 10th May. To protect him from the rudeness of the learned and unlearned public (scientific and unscientific people), as well as of the press, and to prevent a possible repetition here of his expulsion by the police * at the demand of the public, we had taken care, as at his second visit in December of last year, wholly to seclude him from the public.

As regards the following experiments with Mr. Slade, I describe them in the first place for *physicists*, that is, for scientific men who are competent to understand my other physical investigations and experiments, to which, during the space of twenty years, I have given publicity in scientific journals. Such men alone are able to form an independent judgment, on the ground of my antecedent work, as to how far confidence should be extended to me as a physical experimentalist. For though the theoretical considerations—by which the facts of observation so imparted by me during that space have been connected hitherto—deviate in many respects from my own, the facts themselves so observed by me have up to this time received only confirmation in their entirety. As regards such men, also, who on the ground of my labours heretofore are able to form their own independent judgment on my reliability and

* That had happened at Vienna.—Tr.

credibility, I am relieved from the useless trouble of describing more minutely and circumstantially than is necessary for intellectual and scientific men, the conditions under which the following phenomena were observed by me. Suppose, for example, I observed during a physical investigation (as in that concerning the electric fluid) deviations of the magnetic needle under hitherto unusual conditions. If now a physicist, wishing to bring my observations into contempt, were to suggest that I had perhaps accidentally had a magnetic knife on the table, or had not duly taken into account the daily variations of the earth's magnetism, such suppositions might be entertained with respect to a student or beginner in the province of physical observations, but I myself should feel them, coming from a scientific colleague, as an insult, and should hold it beneath my dignity as a physicist to reply to them.*

I assume entirely the same position in describing the following experiments with Mr. Slade, which I

---

* The above protest recalls that of Mr. Crookes, in referring to a suggestion that, in his researches with Mr. Home, he had possibly allowed the latter to supply a board forming an essential part of the apparatus employed.

"Is it seriously expected," says Mr. Crookes, "that I should answer such a question as 'did Mr. Home furnish the board?' Will not my critics give me credit for the possession of some amount of common sense? And can they not imagine that obvious precautions, which occur to them as soon as they sit down to pick holes in my experiments, are not unlikely to have also occurred to me in the course of prolonged and patient investigation?"—TR.

conducted partly alone, partly in company with my above-named friend Oscar von Hoffmann, as in describing the greater number of my former physical investigations.

With respect to the preposterous demand, on entering a new, and to us wholly unfamiliar, province of physical phenomena, to impose *à priori* conditions under which these phenomena "*ought*" to occur, I refer to the strictures contained in the above letter of Slade, and in the previous remarks of Herr Aksakow to Herr Geheimrath Virchow at Berlin on the first principles of exact investigation. After this necessary preface I pass on to describe some experiments which I had devised with a view to the confirmation of my space-theory.

The experiments formerly described (17th December 1878) with the knotted cord suggest two explanations, according as one supposes a space of three or of four dimensions. In the first case there must have been a so-called passage of matter through matter; or, in other words, the molecules of which the cord consists must have been separated in certain places, and then, after the other portion of cord had been passed through, again united in the same position as at first. In the second case, the manipulation of the flexible cord being, according to my theory, subject to the laws of a four-dimensional region of space, such a separation and re-union of molecules would not be necessary. The cord would, however, certainly undergo during the process an amount of twisting which would be discernible after the knots were tied. I had not paid attention to this circumstance in December last year, and had not examined the cords with regard to the size and direction of the twist. The following experiment, however, which took place on the 8th of May this year, in a sitting of a quarter of an hour's duration with Mr. Slade in a well-lighted room, furnishes an answer to the above question in favour of the four-dimensional theory without separation of material particles.

The experiment was as follows :—I took two bands cut out of soft leather, 44 centimètres long, and from 5 to 10 millimètres broad, and fastened the ends of each together, as formerly described with the cords, and sealed them with my own seal. The two leather bands were laid separately on the card-table at which we sat; the seats were placed opposite to one another, and I held my hands over the bands (as shown on Plate II.) Slade sat at my left side, and placed his right hand gently over mine, I being able to feel the leather underneath all the time. Slade asserted that he saw lights emanating from my hands, and could feel a cool wind over them. I felt the latter, but could not see the lights. Presently, while I still distinctly felt the cool breeze, and Slade's hands were not touching mine, but were removed from them about two or three

F

decimètres, I felt a movement of the leather bands under my hands. Then came three raps in the table, and on removing my hands the two leather bands were knotted together. The twisting of the leather is distinctly seen in Plate II (copied from a photograph.) The time that the bands were under my hands was at most three minutes. A pair of unconnected strips of leather are also represented on the Plate for clearness of apprehension.

Much pleased, I examined the connected strips of leather for a long time with my friends. I then took a slate myself, and held it with my right hand under the table, in order to repeat the experiment which had succeeded with the Grand Duke Constantine of Russia.* While now, as I did so, Slade's hands, continually visible to me, lay quietly on the table, there appeared suddenly a large hand close in front of me, emerging from under the edge of the table. All the fingers of the hand moved quickly, and I was able to observe them accurately during a space of at least two minutes. The colour of the hand was pale and inclined to an olive-green. And now while I continually saw Slade's hands lying before me on the table, and he himself sat at the table on my left, the above-mentioned hand rose suddenly as quick as an arrow, still higher, and grasped with a powerful pressure my left upper-arm for over a minute long. As

* *Ante*, p. 43.

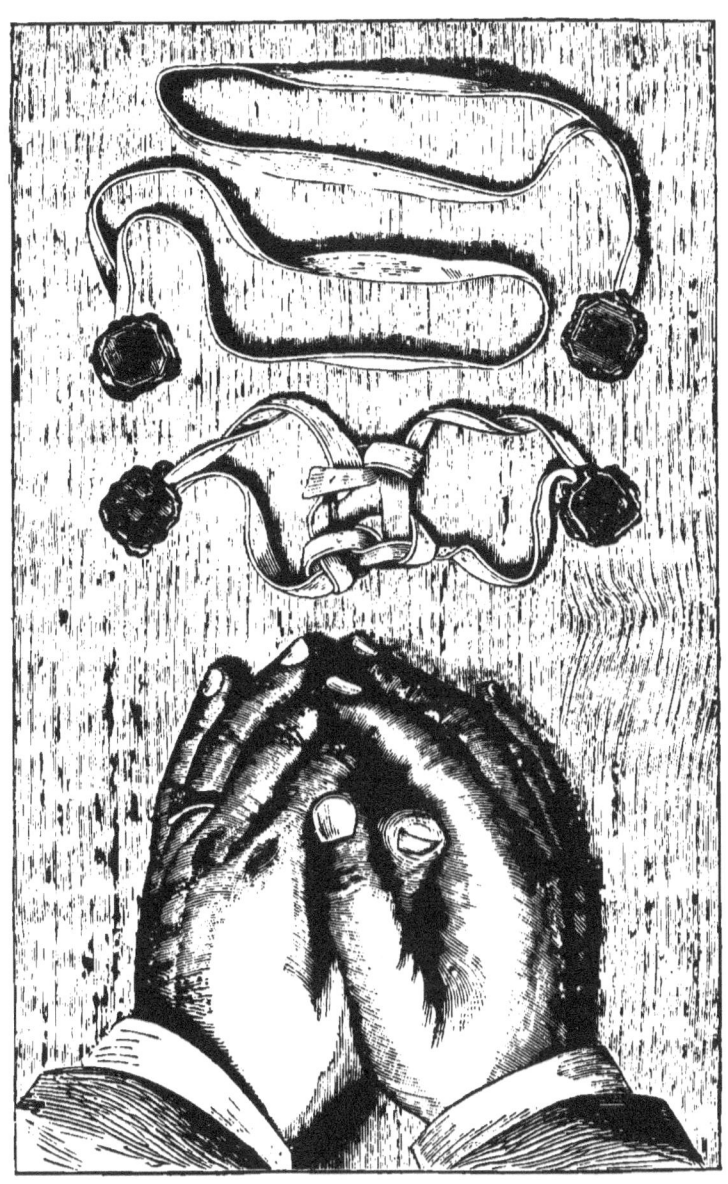

PLATE II.

(Copied from a Photograph.)

my attention was wholly occupied in the observation of the strange hand, and the grip upon my left upper arm happened so suddenly, forcibly, and unexpectedly, I am not able to say anything concerning the condition of the arm which connected the hand with the edge of the table. When this hand had disappeared —Slade's hands lying on the table after as before— I was so violently pinched on my right hand, which during these four minutes was all along holding the above-mentioned slate under the table, that I could not help crying out. With this manifestation the extraordinary sitting closed.

To complete the account of the phenomena of visible and tangible human hands which occurred the year before in presence of my friends and colleagues, Fechner, W. Weber, and Scheibner, I may mention in addition that on the morning of the 15th December 1877, at half-past ten o'clock, while W. Weber and I were again engaged with Slade in the above-mentioned magnetic experiments, suddenly Weber's coat was unbuttoned under the table, his gold watch was taken from his waistcoat pocket, and was placed gently in his right hand, as he held it under the table. During this proceeding, which occupied about three minutes, and was described exactly in its particular phases by Weber, Mr. Slade's hands were, be it understood, before our eyes upon the table, and his legs crossed sideways in such a position that any employment of them was out of the question. The sitting took place in my residence, in the corner room lighted by four large windows, as already described.

Those who seek to explain the phenomena described here, and proved also at other places by reliable observers, of visible and tangible human limbs, by suppositions of possible deception by means of gutta-percha hands, and so forth, treat the matter without consideration, since they judge of phenomena which they have neither seen nor examined referably to the conditions of their occurrence. That such visible and tangible human limbs can, under suitable circumstances, leave behind visible impressions, as, for instance, on flour or sooted paper, will no longer appear surprising after the last-mentioned facts.*

Should the foregoing experiments have afforded proof that there are, outside our perceptible world of three dimensions, things furnished with all the attributes of corporeity which can appear in three-dimensional space and then vanish therefrom, without our being able, from the standpoint of our present space-perception, to answer the questions whence they come and whither they go, then should the following experiment complete this proof, by establishing the appearance and disappearance of bodies which do, in

---

* I may here call attention to the results obtained in London by one of our countrymen, Herr Christian Reimers, and published in "*Psychische Studien*," which results, obtained partly in the presence of Mr. Alfred Russel Wallace, justify the boldest expectations for the future.

fact, belong to our three-dimensional world of space. I have already mentioned (p. 38) the disappearance and reappearance of a small cardboard thermometer-case, and also (p. 53) the sudden appearance of a piece of coal and of wood at a particular place where these bodies had not previously been. Similar and almost more surprising phenomena happened during Slade's residence at Vienna. Baron Von Hellenbach writes me as follows:—

"The disappearance of the book was only superficially treated in my pamphlet,* since therein I only concerned myself with those occurrences which took place beyond the reach of Slade's limbs, as I wished to meet the thoughtless objection, "He did it somehow." The thing happened in the following manner: Slade laid a book and a bit of pencil (at a spot exactly marked) on the slate, which he then conveyed under the surface of the table. The book vanished, and having often been looked for everywhere, fell several times from the ceiling of the room upon the table between the globes of the three-branch chandelier. Once it struck the chain off the roller by which the chandelier was drawn up. A projection by the hand under the table is altogether impossible, since a projected book cannot describe this curve. Slade's upper

* "*Mr. Slade's Residence in Vienna.* An open letter to my friends." (Anonym.), Vienna. Printed and published by T. C. Fischer & Co., 1878. Compare also "*Individualism in the Light of the Biology and Philosophy of the Present*," by Lazar B. Hellenbach, Vienna, 1878 (Braumüller).

and under arm were visible and quiet, and a projection by the foot would as certainly have been remarked as the rise of the book. The experiment was too often repeated, and our attention was too great. I regard as very important a demonstration on your part of a similar disappearance; for if the seen and felt ascent of the slate at my foot proves an unperceived *mechanical* agency, and the production of knots in the endless cord a *four-dimensional* agency, so would the entrance and exit of an object prove another space-dimension, as it were in our immediate neighbourhood, in so stupendous a manner, that it could not be for a moment doubted in my opinion, which is that our illusion of consciousness is nothing but a three-dimensional intuition of a more-dimensional world, brought about by a strange organism. Should your endeavours be similarly successful, I beg you kindly to inform me. B. HELLENBACH."

I had received the above letter at eight in the morning of the 5th May. Without having mentioned it to Slade or to Herr O. von Hoffmann, I expressed the wish, at the sitting which took place with Mr. Slade at eleven o'clock, to have the opportunity of observing again, as in December of the year before, the disappearance and reappearance of a material body in some very striking manner. Ready at once for the experiment, Slade requested Herr von Hoffmann to give him a book; the latter thereupon took from the small bookshelf at the wall a book printed and bound in octavo. Slade laid this upon a slate, held the same partly under the edge of the table, and immediately withdrew the slate again *without the book.* We searched the card-table carefully everywhere, outside and inside. So also we searched the small room, but all in vain; the book had vanished. After about five minutes we again took our places at the table for the purpose of further observations; Slade opposite me, Von Hoffmann between us on my left. We had scarcely sat down when the book fell from the ceiling of the room on to the table, striking my right ear with some violence in its descent. The direction in which it came down from above seemed from this to have been an oblique one, proceeding from a point above and behind my back. Slade, during this occurrence, was sitting in front of me, and keeping both his hands quietly on the table. He asserted shortly before, as usual on occasions of similar physical phenomena, that he saw lights either hovering in the air or attached to bodies, whereof, however, neither my friend nor myself were ever able to perceive anything.

In the sitting of the following day, the 6th May, at a quarter-past eleven, by bright sunshine, 1 was to be witness, quite unexpectedly and unpreparedly, of a yet far more magnificent phenomenon of this kind.

I had, as usual, taken my place with Slade at the

card-table. Opposite to me stood, as was often the case in other experiments, a small round table near the card-table, exactly in the position shown in the photograph (taken from nature) upon Plate III [see page 106], illustrating the further experiments to be described below. The height of the round table is 77 centimètres, diameter of the surface 46 centimètres, the material birchen-wood, and the weight of the whole table 4·5 kilogrammes. About a minute might have passed after Slade and I had sat down and laid our hands joined together on the table, when the round table was set in slow oscillations, which we could both clearly perceive in the top of the round table rising above the card-table, while its lower part was concealed from view by the top of the card-table.

The motions very soon became greater, and the whole table approaching the card-table laid itself under the latter, with its three feet turned towards me. Neither I nor, as it seemed, Mr. Slade, knew how the phenomenon would further develop,* since during the space of a minute which now elapsed nothing whatever occurred. Slade was about to take slate and pencil to ask his "spirits" whether we had anything still to expect, when I wished to take a nearer view of the position of the round table lying, as I supposed, under

---

* The movement of heavy objects without any possible contact by Slade was so common that we looked on the movement of the table as only the beginning of a further succession of phenomena.

the card-table. To my and Slade's great astonishment we found the space beneath the card-table completely empty, nor were we able to find in all the rest of the room that table which only a minute before was present to our senses. In the expectation of its reappearance we sat again at the card-table, Slade close to me, at the same angle of the table opposite that near which the round table had stood before. We might have sat about five or six minutes in intense expectation of what should come, when suddenly Slade again asserted that he saw lights in the air. Although I, as usual, could perceive nothing whatever of the kind, I yet followed involuntarily with my gaze the directions to which Slade turned his head, during all which time our hands remained constantly on the table, linked together (*über-einander liegend*); under the table, my left leg was almost continually touching Slade's right in its whole extent, which was quite without design, and owing to our proximity at the same corner of the table. Looking up in the air, eagerly and astonished, in different directions, Slade asked me if I did not perceive the great lights. I answered decidedly in the negative; but as I turned my head, following Slade's gaze up to the ceiling of the room behind my back, I suddenly observed, at a height of about five feet, the hitherto invisible table with its legs turned upwards very quickly floating down in the air upon the top of the card-table. Although we involuntarily drew back our heads sideways, Slade to the left and I to the right, to avoid injury from the falling table, yet we were both, before the round table had laid itself down on the top of the card-table, so violently struck on the side of the head, that I felt the pain on the left of mine fully four hours after this occurrence, which took place at about half-past eleven.

Chapter Sixth.

THEORETICAL CONSIDERATIONS—PROJECTED EXPERIMENTS FOR PROOF OF THE FOURTH DIMENSION—THE UNEXPECTED IN NATURE AND LIFE—SCHOPENHAUER'S "TRANSCENDENT FATE."

THE foregoing facts of observation are thus *empirically* at variance with the dogma of the unchangeableness of the quantity of matter in our *three-dimensional* world.* Since, however, that dogma of the constancy of substance cannot derive its *dogmatic* character from experience, but merely from the principles of our *reason*, which are inherent in our mind just as in the *à priori* law of causality, that is to say, *before* all experience: there is thus imposed on our *reason* the task of freeing our *understanding* from the above contradiction between the facts of observation and a principle of our reason. I have already shown in detail, in the first volume of these treatises, how very easily this problem is solved by the acceptance of a fourth dimension of space. The table which dis-

* Not altogether: as it might be suggested that the vanished objects only assumed a gaseous form—the quantity of matter thus remaining the same, as in the case of combustion.—TR.

appeared during six minutes must nevertheless have existed *somewhere*, and the quantity of the substance constituting it must, according to the above principle of reason, have remained absolutely constant. If, however, we can only answer the question " where ? " by assigning a place—and it has been empirically shown that this place cannot lie in the region of space of three dimensions perceptible to us,—it follows of necessity that the answering of the question " where ? " hitherto so easy to us, must be an *incomplete* answer, and therefore one both requiring and capable of amplification. How by this means also the conception of juxtaposition obtains an extension by help of the fourth dimension of *absolute* space, I have already above explained at length in a note,[*] to which I may refer my readers.

So also I have already shown in the treatise "On Action at a Distance," vol. i. p. 269, that the so fruitful "Axiom of the Conservation of Energy" retains all its validity for space of four dimensions, while at another place I remarked, " If one regards the distance of two atoms and the intensity of their interaction, in our three-dimensional space, as projections of similar magnitudes from a space of four dimensions, a change would be effected in the magnitudes, form, and supply of kinetic energy of the three-dimensional projection (the material body), simply through alterations in the

[*] *Ante*, p. 58.

relative positions of the four-dimensional object, without these properties in the latter undergoing any change. *The axiom of the conservation of a constant amount of energy thus retains its full validity for space of four dimensions, nay, on closer consideration, it is even the premiss on which rests the correspondence of the extended conceptions of space to physical occurrences.*" *

To the considerations offered in the early part of this treatise concerning the "actual" or "real" lying at the ground of space, I may here add the following words of Riemann : †—

" The question of the validity of the postulates of geometry in the infinitely little is connected with the question of the inner principles of the mass-relations of space. In this question, which can well be accounted as still belonging to the doctrine of space, the above observation has the application that in a discrete diversity (*Mannigfaltigkeit*) the principle of mass-relations is already contained in the conception of this diversity, whereas, in a continuous diversity, this principle must come to it from without (*anders woher hinzukommen muss*). Thus, either the reality underlying space must form a discrete diversity, or the principle of mass-relations must be sought *without*

---

† Riemann's collected mathematical and posthumous scientific works, edited, with the assistance of R. Dedekind, by H. Weber, Leipsic : (Teubner), 1876.

(*ausserhalb* *), in binding forces acting thereon (*in darauf wirkenden bindenden Kräften*).

"The decision of these questions can only be found by transcending the hitherto empirical conception of phenomena, of which Newton established the principle, and, impelled by *facts which cannot be explained by it*, gradually reforming this conception. Such researches, which, like the present, transcend common conceptions, can only serve to prevent *this work being hindered by the narrowness of ideas, and advance in knowledge of the connection of things being impeded by traditional prejudices*. This carries us over into the province of another science, that of Physics, which is not permitted by the nature of our present subject."

These words of Riemann prove incontrovertibly that he, as one of those acute founders of the theory of an *extended* space-conception, recognised as thoroughly necessary the introduction (*Hinzuziehung*) of *physical* elements (*Momente*); that is, derived from *observed* facts.†

---

* The word "*ausserhalb*" in relation to the whole circuit of the three-dimensional region of space given perceptibly to us has only one sense, if for the centre of those "binding forces, acting thereon" is presupposed a fourth dimension.

† The recently introduced "conception of solidity" or "rigidity" (*der Festigkeit oder Starrheit*) is only another expression for this physical side of the problem. For though the geometrical conception of solidity can be *defined* as the unchangeableness of the distance of the points of a system of points, yet the intuition underlying this "conception" is only derived from experience, just as the conception of motion is abstracted from experience. Compare Helmholtz "On the Origin and Meaning of Geometrical Axioms" (*Popular Scientific Essays*, November 3rd, 1876). So

I now proceed to the description of further successful experiments in the presence of Mr. Slade, which will partly confirm those already mentioned, partly establish them more thoroughly by new modifications.

In order to exclude as far as possible the dependence of to us inexplicable phenomena upon human testimony, I desired to devise experiments such that the permanent effect, as final result, should be completely unexplainable according to the conceptions we have hitherto entertained of the laws of nature. With this object, I had arranged the following experiment:

1. Two wooden rings, one of oak, the other of alderwood, were each turned from one piece.* The outer diameter of the rings was 105 millimètres, the inner 74 millimètres. Could these two rings be interlinked without solution of continuity, the test would be additionally convincing by close microscopic examination of the unbroken continuity of the fibre. Two different kinds of wood being chosen, the possibility of cutting both rings from the same piece is likewise excluded. Two such interlinked rings would consequently in themselves represent a "miracle," that is, a phenomenon which our conceptions heretofore of

---

Wilhelm Fiedler, "Geometry and Geomechanics," in the "Fourth Yearly *Journal* of the Society of Natural Philosophy at Zurich," 21st yearly vol., 1876, same number "On Symmetry" by Fiedler, number 2, p. 186 *et seq.*

* Both these rings I received in February of this year, through the kindness of Herr G. De Liagre. I take this opportunity publicly to thank this gentleman, as also the frequently-mentioned Herr Oscar von Hoffmann, for their energetic assistance in the experiments with Mr. Slade.

G

physical and organic processes would be absolutely incompetent to explain.

2. Since among products of nature, the disposition of whose parts is according to a particular direction, as with snail-shells twisted right or left, this disposition can be reversed by a four-dimensional twisting of the object, I had provided myself with a large number of such shells, of different species, and at least two of each kind.

3. From a dried gut, such as is used in twine-factories, a band without ends (*in sich geschlossenes*) was cut, of a breadth of from four to five millimètres, and a circuit of 400 millimètres. Should a knot be tied in this band, close microscopic examination would also reveal whether the connection of the parts of this strip had been severed or not.

4. In order to demonstrate yet more evidently the so-called penetration of matter, which comes in question in all these experiments, I had a glass ball, enclosed on all sides, of 40 millimètres diameter, blown by the glass-manufacturer, Herr Gotze, of this place. From a paraffin candle I had then cut off with a sharp knife a piece of such a length that it just fell short of that of the interior of the ball. I asked Herr Gotze if he thought it possible to blow a glass ball of the prescribed size round such a piece of paraffin provided with sharp edges, without melting the paraffin, at least at the edges. He replied most decidedly in the nega-

tive; and even independently of his authority, I believe I do not risk contradiction in asserting that such a piece of paraffin with *sharp* unmolten edges in the interior of the said glass ball would be, according to our heretofore *limited* conception of the laws of nature, an inexplicable miracle.

The foregoing preparations sufficiently show what sort of phenomena I *wished* to see in Slade's presence. Since, however, in the course of more than thirty sittings with Mr. Slade, I had come to the conviction that he did not himself "do" the mysterious things which happened near him, I could not rationally demand of him that he should "show" me all the above-mentioned experiments. Far more unreasonable still must I have hence considered the desire on my part to impose "conditions" on Mr. Slade, under which he should effect these to himself inexplicable proceedings. I preferred therefore to comport myself towards Mr. Slade and the phenomena occurring in his presence just as I did towards nature in my physical discoveries up to that time, or to the previously anticipated fall of meteors, which happened when our earth crossed the path of Biela's comet, on the 27th November 1872. I accordingly remained patient, and in a passive receptive disposition for the things which should come, and left it confidently to nature of her own free-will to reveal to me as much of her secrets as seemed fitting to her without blinding my intellectual eyes by the splendour of her majesty; mindful always of Goethe's words :—

"*Geheimnissvoll aus lichten Tag
Lässt sich Natur des Schleiers nicht berauben,
Und was sie deinem Geist nicht offenbaren mag,
Das Zwingst du ihr nicht ab mit Hebeln und mit Schrauben.*" *

" Inscrutable in noon-day's blaze,
Nature lets no one tear the veil away;
And what herself she does not choose
Unasked before your soul to lay,
You shall not wrest from her by levers or by screws."
—*Theodore Martin's translation.*

And in fact, I know no better comparison whereby to indicate the character of the constantly *unexpected* occurrences in their succession and ingenious connection, than the manner in which men are led by fate. Seldom happens just that which we, according to the measure of our limited understanding, wish; but if, looking back on the course of some years, we regard

---

* Faust alone, after Wagner had left him with the words "*Zwar weiss ich viel, doch möcht ich alles wissen*" (much, it is true, I know; yet would know all). In the monologue that follows, Faust expresses his sentiments upon that "Famulus," and doubtless later "professors," in these words :—

"*Wie nur dem Kopf nicht alle Hoffnung schwindet,
Der immerfort an schalem Zeuge klebt;
Mit gier'ger Hand nach Schätzen gräbt,
Und froh ist, wenn er Regenwürmer findet.*"

" Strange that all hope has not long since been blighted,
In one content on such mere chaff to feed;
Who digs for treasure with a miser's greed,
And if he finds a muck-worm is delighted."
—*Theodore Martin's translation.*

what has actually come to pass, we recognise gratefully the intellectual superiority of that Hand which, according to a sensible plan, conducts our fates to the true welfare of our moral nature, and shapes our life dramatically to a harmonic whole. *Volentem fata ducunt, nolentem trahunt*, says an old proverb, often quoted by Schopenhauer. That such a conception of the significance, and of the inner intellectual connection of our fate, does not merely spring from an idealism coloured by optimism, but powerfully imposes itself even on a pessimist with sufficiently high powers of understanding, we have the most striking proof in Schopenhauer's treatise, " On apparent Design in the Fate of the Individual" (*Uber die anscheinende Absichlichkeit im Schicksale des Einzelnen*). He says : *

"At all events, however, the perception, or rather the opinion, that this necessity of all that happens is no blind necessity, thus the belief in an evolution of the events of life not less methodical than necessary, is a fatalism of a higher kind, though not so easily demonstrable, and one which perhaps occurs to every one, sooner or later, at one time or another, and is held by him, for a time, or ever after, according to his mode of thinking. We might name it *transcendent fatalism*, to distinguish it from the common and *demonstrable* fatalism. . . . Thus, in regard to particular individual fate, grew up in many that *tran-*

* Parerga and Paralipomena, vol. i. pp. 218, 219.

scendent fatalism, which the attentive consideration of his own life, after its thread has been spun to a considerable length, suggests, perhaps, to every one once; and which has not only much that is consolatory, but it may be also much that is true; and therefore has it at all times been affirmed, even as dogma. Neither our conduct nor our career is *our* work; but that, indeed, which nobody supposes to be so—our *nature* and *existence* (*unser Wesen und Dasein*). For on the foundation of these, and of the circumstances and external events occurring in the strictest causal connection, our actions and whole career proceed with complete necessity. Already at a man's birth, therefore, is his whole career irrevocably determined even in its details, so that a somnambule in high power could predict it exactly. We should keep this great and certain truth in view in the consideration and judgment of our career, our acts and sufferings."

Chapter Seventh.

VARIOUS INSTANCES OF THE SO-CALLED PASSAGE OF MATTER THROUGH MATTER.

AFTER this digression, I now go on to the description of those physical modifications which have actually been effected in some of the above objects prepared by me, *without their having been touched at all by Slade.*

On the 3rd May of this year at half-past eight in the evening, during a sitting in which, besides myself, Herr O. von Hoffmann took part, there lay on the table with other objects two of the above-mentioned snail-shells. I had bought both of them on the morning of the same day from an Italian shell-dealer, who offered his wares for sale at Leipzig fair. The smaller shell belonged to a species commonly found here; the larger to a species which, according to the dealer, is found on the shore of the Mediterranean Sea; he wrote down the name of it—*Capo Turbus* (Lat. *Caput turbo*)—at my desire. The nearly circular aperture of this shell had a diameter of about 43 millimètres, while the smaller one measured only about 32 millimètres in its greatest extent. On this evening I had, without definite design, so capped the smaller shell with the larger, that the latter, lying with its opening next the surface of the table, completely hid the former. This had happened during a sitting in which wholly different manifestations occurred. When, now, Slade held a slate * under the edge of the table in the usual way to get writing on it, something clattered suddenly on the slate, as if a hard body had fallen on it. When immediately afterwards the slate was taken out for examination, there lay upon it the smaller shell which a minute before I had capped with the larger, as above mentioned. Since both shells had lain before almost exactly in the middle of the table, untouched and constantly watched by me, here was, therefore, the often observed phenomenon of the so-called penetration of matter confirmed by a surprising and quite unexpected physical fact. Reserving the account of numerous other phenomena of this kind to the third volume of my *Scientific Treatises*,† I yet mention here one very remarkable circumstance. Immediately after Mr. Slade drew the slate from under the table with the smaller shell on it, I seized the shell in order closely to examine it for any changes that might have happened in it. I was nearly letting it drop, so very

---

* In order to deprive the suggestion, that Mr. Slade writes himself on the slate, by means of a bit of pencil inserted under the finger-nails, of every *rational* foundation, I had provided myself, from the stationery establishment of Mylius at this place, with half-a-dozen slates having a length of 34 centimètres and a breadth of 15 centimètres (with the fabric mark, A. W. Faber, no. 39). With a slate so much longer than usual, it was impossible that Mr. Slade could write with his fingers, while holding the slate, over its whole surface.

† Post.

hot had it become. I handed it at once to my friend, and he confirmed the fact of its remarkably high temperature. This fact is, I believe, of physical importance with regard to one circumstance in the following experiments.

On the 9th May, at seven o'clock in the evening, I was alone with Slade in our usual sitting-room. A fresh wind having blown all the afternoon, the sky was remarkably clear, and the room, which has a westerly aspect, was brilliantly lighted by the setting sun. The two wooden rings and the above-mentioned (p. 98) entire bladder band were strung on to a piece of catgut one millimètre in thickness, and 1·05 metre in length. The two ends of the catgut were tied together by myself in a knot, and then, as formerly in the case of the string, secured with my own seal by myself. Plate III. represents the condition of things at the beginning of the sitting; Plate IV., at its conclusion.

When Slade and I were seated at the table in the usual manner, I placed my two hands over the upper end of the sealed catgut, as shown in the plate, photographed from life. The small round table, already referred to, was placed shortly after our entry into the room, in the position shown in the picture.*

* It is scarcely necessary to remark that the photographs were taken, not during, but some days after, the sittings. The two tables are those used in the sittings, but the sealed catgut, with the two wooden rings and the strip of bladder, were afterwards prepared to show the condition of these objects before the sitting, and are as far as possible exactly copied from the originals shown on Plate III.

After a few minutes had elapsed, and Slade had asserted, as usual during physical manifestations, that he saw lights, a slight smell of burning was apparent in the room—it seemed to come from under the table, and somewhat recalled the smell of sulphuric acid. Shortly afterwards we heard a rattling sound at the small round table opposite, as of pieces of wood knocking together. When I asked whether we should close the sitting, the rattling was repeated three times consecutively. We then left our seats, in order that we might ascertain the cause of the rattling at the round table. To our great astonishment we found the two wooden rings, which about six minutes previously were strung on the catgut, in complete preservation, encircling the leg of the small table. The catgut was tied in two loose knots, through which the endless bladder band was hanging *uninjured*, as is seen in Plate IV. [See Plate X., Appendix D.]

Immediately after the sitting, astonished and highly delighted at such a wealth of *permanent* results, I called my friend and his wife into the sitting-room. Slade fell into one of his usual trances, and informed us that the invisible beings surrounding him had endeavoured, according to my wish, to tie some knots in the endless band, but had been obliged to abandon their intention, as the band was in danger of "melting" during the operation under the great increase of temperature, and that we should perceive this by the white-

PLATE III.

(*Copied from a Photograph.*)

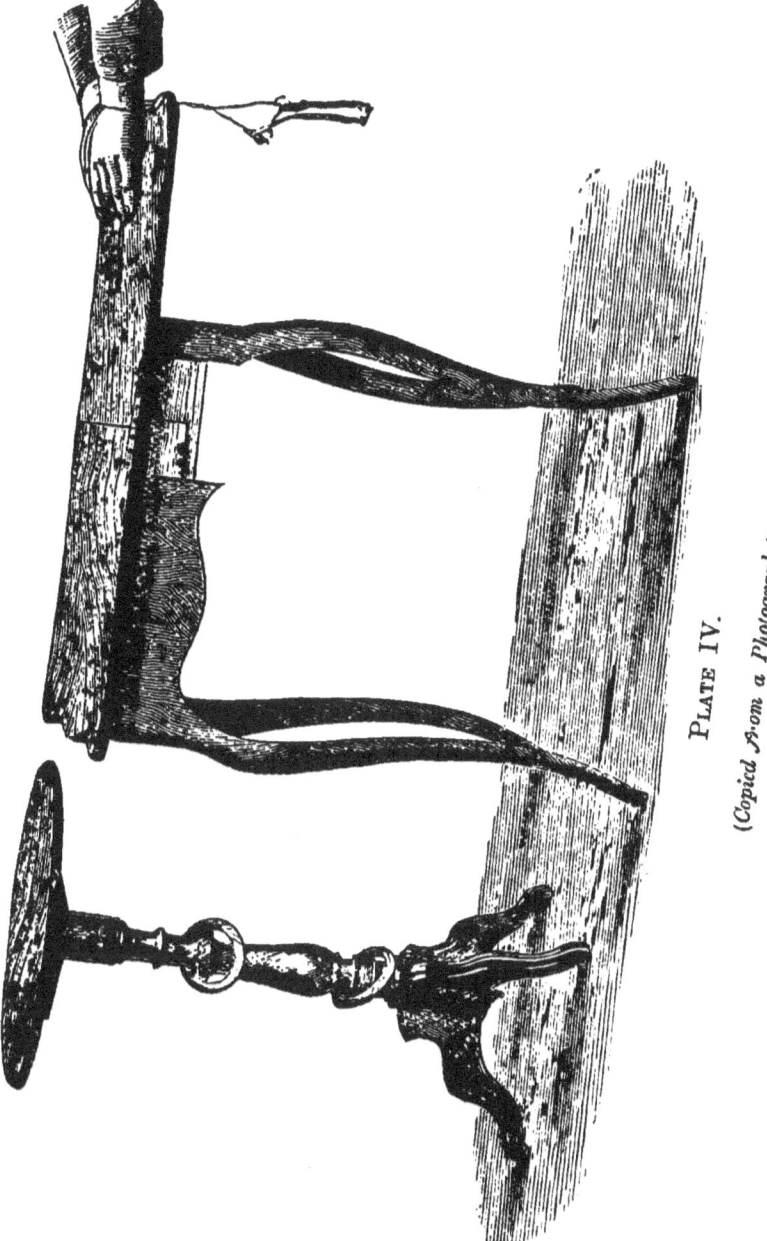

PLATE IV.

(Copied from a Photograph.)

Plate V.

(*Copied from a Photograph.*)

ness of a spot on the band. Having taken the band into my own hands immediately after the sitting, and held it up to the moment of Slade's communication, I felt great interest in testing the correctness of this assertion. There was, in fact, a white spot as indicated, and when we took *another* piece of exactly the same material and held it over a lighted candle, the effect of the increased temperature was to produce precisely such another white spot. This fact, in connection with the burning smell perceived during the sitting, as well as the increase in temperature in a former experiment (related above), will be worth bearing in mind in further experiments with four-dimensional movements of bodies.

In fact, if, according to the above-cited alternative of Riemann, "the reality underlying space must be sought in binding forces acting thereon," so could such increase of temperature be produced in like manner as in the motions of conducting bodies in the magnetic field. For suppose we knew nothing of the magnetic induction discovered by Faraday, and were observing in a space lying between the poles of an electro-magnet, not otherwise perceptible to us, the increase in temperature of quickly-moved conducting bodies would appear to us just as wonderful and incomprehensible as the heat produced in mundane bodies in the above instances by four dimensional changes of place.

Doubtless, a highly-developed understanding, which from *metaphysical* principles, that is, from principles derived from reason, had recognised the necessity and universal significance of Weber's law for every interaction of spatially separated bodies, would have inferred the existence of Faraday's magnetic induction, *à priori*; he would therefore regard the heating of conducting bodies on their motion only as an empirical confirmation of his *à priori* deductions, and thus would have inferred the real existence of such an electromagnet, even if his mortal eye had never seen it, and his mortal body had never touched it.

From the foregoing it will be seen that my *prepared* experiments did not succeed in the manner *expected* by me. For example, the two wooden rings were not linked together, but instead, were transferred within five minutes from the sealed catgut to the leg of the round birchen table. Since the seal was not loosened, and the top of the table was *not at any time* removed— it is still tightly fastened—it follows, from the standpoint of our present conception of space, that each of the two wooden rings penetrated, first the catgut, and then the birch wood of the leg of the table. If however, I ask whether, in the eyes of a sceptic, the experiment desired by me, or that which actually succeeded, is most fitted to make a great and convincing impression, on closer consideration every one will decide in favour of the latter. For the demonstrative force of the interlinked rings would rest merely on the credibility of the botanically-educated microscopist, who must have been my witness (as the Imperial Court conjuror, Bellachini, was for Mr. Slade), that the natural conformation of the rings had never been disturbed. How wholly useless, however, such testimonies are at present, when, according to Goethe's expression, " incredulity has become like an inverted superstition for the delusion of our time," we have seen in the sort of criticism which Bellachini's testimony has undergone at the hands of the Berlin literati.* The question will moreover be asked, why just here in Leipzic the experiments with Mr. Slade have been crowned with such splendid success, and yet the knot experiment, for example, has not once succeeded in Russia, notwithstanding so many wishes. If it is considered how great an interest Mr. Slade must have in seeing so simple and striking an experiment everywhere and always successful, every rightly judging and unprejudiced person must see just in this very circumstance the most striking proof that Mr. Slade is no trickster who by clever manipulations makes these knots himself. For such an one would evidently be at the trouble so to increase his expertness, by frequent repetition of the experiment, as to be able to rely with certainty on his art to deceive other "men of science." That, never-

* Mere contemptuous abuse—Professor Zöllner gives the articles at length in an earlier part of his volume.—TR.

theless, this obvious consideration has not suggested itself, the above-mentioned failure being regarded, on the contrary, as just the proof that Mr. Slade has only deceived us at Leipsic, which he could not do with the higher intelligence of the Russian learned, is shown by the following words of a scientific friend from Russia, to whom I had sent my "Scientific Treatises."

—— " February 22, 1878.

" Perhaps the following fact may open your eyes. Two days ago, in consequence of your letter and in dependence on it, two scientific friends visited Mr. Slade, and requested him to undertake in their presence the striking operation of the four knots. Mr. Slade's answer was, 'This operation has only succeeded twice (in Leipsic?); at present my medium is not strong enough for it.' After this can you look upon that operation as an actual proof of the existence of the fourth dimension?"

It has further been asked, why the communications which are written for Mr. Slade on his slates, as is supposed by invisible spirits, are for the most part so commonplace, and so completely within the compass of human knowledge; high spirits must yet necessarily write with more genius, and also spell properly. A private teacher of philosophy at Berlin having made this objection to me personally, on his visit to Leipsic, I observed to him that any communication transcending the present horizon of our understanding must necessarily appear to us absurd and incomprehensible, and I quoted to him the following words of Lichtenberg :—* "If an angel were to discourse to us of his philosophy, I believe that many propositions would sound to us like '2 and 2 make 13.'" Far from understanding me, that young philosopher asked me quite seriously, and with an expression of the highest curiosity, whether such propositions, then, ever appeared on Mr. Slade's slates to attest their angelic origin. Completely unprepared for such a naïve question, I was silent, and looked with some astonishment at my young philosopher, who had even already published a book on the new theory of space. Without replying, I thought, "Only wait; soon thou also wilt be at rest" ("*Warte nur, balde ruhest auch du*"), as regular professor of philosophy in the bosom of some famous German university, and then will it be with thy students just as with us "if an angel discoursed to us of his philosophy;" for Lichtenberg says, "We live in a world where one fool makes many fools, but one wise man only a few wise men." †

The fact that, just here in Leipsic, experiments devised from the standpoint of a definite theory have been so surprisingly successful in the presence of Mr. Slade, I regard as one of the most striking proofs of the

* *Miscellaneous Writings*, vol. i. p. 105.
† *Thoughts and Maxims*, p. 46.

great intelligence of the invisible beings surrounding him. For if, without appearing presumptuous, I may include myself in that class of intelligent beings in which indeed all my fellow-men also number themselves, by the name of their species "*homo sapiens,*" yet would I make more precise communications and explanations concerning my physical observations only to such men as I hold to be sufficiently trained. In a society of social democrats, or in one of German or English scientists, where Mr. Tyndall or Sir W. Thomson finds such a ready sale for their wares*—yes, even in the Berlin Academy, I would refrain from speaking or experimenting on my theory of space. Were I, for example, myself one of those invisible spirits who hover round Mr. Slade, and were my medium invited to a "scientific" examination by the Berlin academicians, it would be easy for me to write on the slate the following proposition—for instance, " We are the play of our brain-molecules," or, " The first life on the earth took its rise in germs enclosed in the cool folds of a meteoric stone."

These propositions would evidently have been greeted with joy by Mr. E. du Bois-Raymond and Herr Helmholtz as striking proofs of the high intelligence of those invisible beings, and would certainly have

---

* To make this allusion intelligible, it should be mentioned that much of these volumes is devoted to criticism of the atomic and other speculative hypotheses of these scientific gentlemen.—TR.

brought much honour and glory to my medium. As an invisible spirit, I might perhaps have perpetrated in good-humour such a jest with the Berlin academicians, just as Sir W. Thomson did with his "unscientific people" at the Edinburgh meeting of scientists seven years ago.* Since, however, in the higher world of spirits truth is held as something sacred, with which only lower spirits permit themselves to jest, so by such purport of my slate-writing should I have made myself guilty of an injury to the moral law, which, according to the laws of divine and eternal justice, would bring its own punishment. May not possibly similar considerations have prevailed to hinder Slade's invisible beings from displaying at another place their treasures, which have been shown to us partly here in Leipzic in such wonderful abundance?

Lastly, a circumstance may be briefly noticed which relates not so much to the moral and intellectual qualities of the invisible spirits as to those of the visible mediums, whom those spirits need for their manifesta-

* When he made the suggestion that the first life on this earth originated in germs enclosed in meteorites. This idea was for a long time discussed quite seriously by, among other scientific authorities, E. du Bois Raymond, Helmholtz (who claimed priority of it for himself, and by Zöllner. But in "*Nature*" of 4th July 1874 appeared the following, in a criticism of Zöllner's book, "*On the Nature of Comets:*"—"The celebrated moss-grown fragments from the ruins of another world was only a jest, taken in earnest even by many of our own countrymen, so we can scarcely reproach Professor Zöllner for falling into the same mistake."—TR.

tions. It has been alleged as a characteristic of *all* such mediums, that notwithstanding the most wonderful occurrences in their proximity, they have yet the inclination to deceive, that is, when opportunity offers, to produce the desired effect by such operations as they consciously endeavour to hide from observation. Having regard to the great danger of such attempts to the medium, and to the entire disproportion between the effects which can be so produced by an inexperienced trickster and those resulting from genuine mediumship, the question arises whether, when this is the case with a medium who has been proved with certainty to be really such, the same consideration does not apply as with persons suffering under so-called kleptomania? It is asserted that a well-known and highly-gifted lady in distinguished circles of Berlin society suffers from this disease. For example, after making large purchases at a jeweller's shop, she will secretly abstract an ornament, which, when she has got home, she will return by her servants to the proprietor. Sometimes a similar perversion of the moral instinct appears with women in the state of pregnancy. In all these cases we do not hold the persons in question morally accountable for these proceedings, since the end attained thereby is out of all proportion, considering the innocent and suitable means at hand. Although I never, during my thirty sittings and other intercourse with Mr. Slade, perceived anything of such perverse methods, yet I ask every unprejudiced person whether, if this has been the case elsewhere, the above morally and legally admissible judgment in relation to kleptomaniacs is not here also exculpatory, considering the certainly anomalous physiological constitution of such mediums.

Reserving till later on in these treatises the detailed communication of further and not less remarkable phenomena which happened in Slade's presence, I will here add an observation to the accurate description (*supra*, p. 34) of the physical manifestation which occurred on the occasion of Slade's first visit, on the 16th November 1877, in my house and in the presence of my friends and colleagues, Wilhelm Weber and Scheibner.[*] In all phenomena in the presence of spiritualistic mediums hitherto observed and published, it is almost exclusively the *modus operandi* that has led to controversies concerning the explicability of the phenomena from the standpoint of our conception of nature heretofore. An argument has been founded on the fact that things occur also in the presence of conjurers, in which the *modus operandi* of the performer is concealed from us, and thus the causal connection between the muscular movements of the artist and the effect produced by him is so interrupted (apparently), that for the spectator there arises

[*] The sudden rending of the wooden frame of a bed-screen at least five feet from Slade.

the impression of the inexplicable, and therefore of the miraculous. This argument, however, has for its premiss the understood and thus unexpressed presupposition that the muscular force requisite for the production of these tricks of the conjurer remains *within* those limits which according to experience are prescribed to human beings by the organisation of their bodies.

If, for example, one man alone were to perform a trick requiring the strength of two horses, in relation to such a result the above argument would be no longer admissible, since then there would be no conceivable *modus operandi* able to produce the effect.

In the case of my bed-screen—the manifestation mentioned at p. 34—I am fortunately able to establish such an instance.

The material of the frame was alder wood; the screen was new, and had been bought by me about a year before at the furniture shop already mentioned (p. 36). The cross-cut of the two pieces of wood which were longitudinally * and simultaneously rent, above and below, amounted to 3·142 cubic centimètres.

---

* That the pull (*zug*) upon the screen has in fact acted longitudinally only is still evidenced quite independently of the above-mentioned direction of the fibres at the places of division (p. 34). For between the two strong beams for connecting the movable parts of the frame are two thin, parallel pieces of wood for securing the green, woollen stuff with which the screen is overlaid. These thin pieces are fastened without glue to the vertical supports loosely in holes about 25 millimètres deep; if, therefore, instead of a longitudinal pull, a rupture (*bruch*) had taken place, these two pegs must have been broken away, which was not the case.

According to the experiments of Ettelwein,* the amount of pull requisite for the longitudinal rending of such a piece of alder wood is 4957 kilogrammes, or about 99 cwts; since, therefore, *two* such rods have been simultaneously rent, for the production of this effect a force of pull (*Zugkraft*) amounting to 198 cwts. must have been used.

In order, now, to compare the force here given with that exercised by men, in what follows I quote literally the appended information from Gehler's *Dictionary of Physics*, vol. ii. p. 976 :—

"The muscles of the thigh hold upright the body, whose weight can be put at 150 lbs.; and since there are muscles which bear 300 lbs. in addition, the weight of pressure already amounts in itself to 450 lbs. To cite, however, some examples only of extraordinary strength, I have myself known a man who without preparation and on an accidental occasion carried six Rhenish cubic feet (Brunswick bushels) of wheat, and upon this a large, strong man, up a flight of about eight steps. This weight of itself can be estimated at 450 lbs. and, with the added weight of the bearer, in the whole at 600 lbs. resting on the feet and legs of that man.

"There are, moreover, many instances of a vastly

---

* *Handbook of Statics of Solid Bodies*, with particular regard to their application to Architecture, vol. iii., Berlin, 1808. A very complete review of earlier experiments is given in the "*Edinburgh Encyclopædia.*" Compare Gehler's *Dictionary of Physics*, vol. ii. p. 138.

greater exertion of strength produced by the extensor muscle of the leg, like that mentioned by Desaguliers, of a man who thus tore a rope which sustained a weight of 1800 lbs. = 18 cwts. ; he himself and some others having raised 1900 lbs. weight by means of a strap hanging down over the hips, by bringing the somewhat bent leg into a straight direction.

"I have myself seen a strong man raise 2000 lbs., by placing himself in a bent posture under a board, whereon this weight rested, bringing its point of gravity somewhere near the hips, supporting the arms on the knees, and then straightening the bent legs. The muscles here applied are, among all in the human body, able to overcome the greatest weights, and so therefore a man raises much heavier burdens in the way described than on the shoulders or with the upper part of the body, if at the same time the backbone has to be straightened.

"I myself knew a man who raised a cwt. from the chair on to the table on the little finger of the right hand with outstretched arm ; and even this instance is by no means the strongest, judging from credible narratives; so I saw the above-mentioned Hercules, who raised the 2000 lbs., grasp with his right hand a perpendicular rod of iron, sufficiently secured, and with outstretched arm keep his whole body sustained in a horizontal position for about five seconds without other support."

Comparing the above with the force of 198 cwts. requisite for the rending of my bed-screen, it will be seen that the strength of the "Hercules" referred to would have to be multiplied by nearly 10—applied in a favourable position—to produce the physical manifestation which took place in Slade's presence without contact. Since "the force in the movement of weights by carrying on the flat" is with a horse on the average about five times greater than that of a man,* so for the production of the mechanical effect in question in Slade's presence, about two horses would have been necessary. Even if Slade should be assumed to be a giant, and the faculty ascribed to him of moving so swiftly in space that my friends Wilhelm Weber, Scheibner, and I myself, were prevented by this rapidity from perceiving how he tore asunder the screen by his own action, yet will *rational* sceptics be disposed to renounce such an "explanation" after the statements just given.

But in case I should be reproached with having in the above supposition caricatured the so-called "rational" attempt at explanation, I may observe that one of my esteemed colleagues who, on the day after the sitting in question, was himself present with two other of our

---

* Gehler's *Dictionary of Physics*, vol. v. p. 1004. Literally "There is, therefore, in the movement by carrying of weights on the flat, a force,
Of a man = 1 according to Coulomb.
Of a horse = 4.8 according to Brunacci.
Of a horse = 6.1 according to Wessermann."

colleagues at a sitting with Mr. Slade, sought quite seriously to appease his scientific conscience by the supposition that Slade carried *dynamite* about with him for the purpose of such strong mechanical manifestations, concealing it in some clever fashion in the furniture, and then with equal adroitness exploding it by a match. This explanation reminded me of one by which a peasant in a remote part of Lower Pomerania attempted to account for the motion of a locomotive. To mitigate in some degree the terror which the first sight of a self-moving locomotive must naturally excite in rude and ignorant men, the priest of the village in question tried to explain to his parishioners the mechanism and effect of a steam-engine. When now the pastor had conducted his peasants, enlightened by this "popular lecture,"* to the railroad just as the first train rushed by, they all shook their heads incredulously, and answered the priest, "No, no, parson, there are horses hidden inside!" That, in fact, within all bodies electrical forces are potentially latent, which, suddenly released, could exceed the strongest effects of a charge of dynamite, I have already remarked in the first volume, as follows: "It is proved that the electrical energy present in the mass of one milligram † of water (or any other body) would be able, if it could be suddenly set free, to

* For reasons given in other parts of his treatises, Professor Zöllner holds popular expositions of scientific subjects in small esteem.—Tr.
† =0.01543 grains.

produce the amount of motion which the explosion of a charge of 16·7 kilogrammes * of powder in the largest of cannons now existing can impart to a shot of 520 kilogrammes."

In the presence of spiritualistic mediums there must therefore have been operative so-called catalytic †

---

* 1 kilogram = lbs. 2.2046213.

† That the ordinary chemical and physical processes require for their explanation the supposition of such catalytic forces was first recognised by Berzelius, with whom, as is well known, the designation of these forces originated.

It is certainly a proof of the great acuteness of Wilhelm Weber, and of the universal significance of his *law*, that already, thirty-two years ago, immediately following the discussion of the analytical expression of his *law* (compare my *Principles of an Electro-Dynamic Theory of Matter*, vol. i.), he expressed himself concerning the existence of catalytic forces in nature as follows:—

"Thus this force depends on the quantity of the masses, on their distance, on their relative velocity, and further on that relative acceleration, which comes to them partly in consequence of the persistence of the motion already present in them, partly in consequence of the forces acting upon them from other bodies."

"It seems to follow from thence, that direct interaction between two electrical masses depends not exclusively upon these masses themselves and their mutual relations, but also in the presence of third bodies. Now it is known that Berzelius has already conjectured such a dependence of direct interaction of two bodies in the presence of a third, and has designated the force thence resulting by the name of catalytic. Adopting this name, it can therefore be said that even electrical phenomena proceed in part from catalytic forces.

"This proof of catalytic forces for electricity is not, however, strictly speaking, a consequence of the discovered principles of electricity. It would only then be so, if with these principles was necessarily connected the idea that only the forces by which electrical masses act *directly* on each other from a distance were thereby determined. It is, however, conceivable that among the forces comprehended under the discovered principles are some exercised *mediately* by electrical masses on one another, which must therefore depend, in the first instance, on the interposing medium,' and furthermore on all bodies acting on this medium. Such mediately exercised forces, if the interposing medium is withdrawn from our view, may easily pass for catalytic forces, although in fact not so. The conception of catalytic forces, hitherto concealed from us, which were able to release and convert into active force a small part of the potential energy laid up in all bodies. That fifty years ago a physicist could venture with impunity publicly to declare the possible existence of " forces, up to the present unknown to us," without on that account having dirt thrown upon him by anonymous writers in (so-called) "respectable" journals, is proved by the following words of the then professor of physics in the University of Heidelberg in the year 1829 :* "Not a few, and among them, moreover, advantageously known scholars, have supposed different *unknown* forces in nature, and especially in man. That there may be such, from whose action many as yet mysterious phenomena of vegetable and animal vital processes could be explicable, certainly cannot be denied generally and *à priori;* but, on the other hand, it is quite certain that the greatest circumspection and a scepticism much to be recommended to a physicist should be exercised in this supposition."

How far the paternal counsel here given to *uncritical* physicists is justifiable and decent when

---

forces must at least be essentially modified in speaking of them in such cases. That is to say, under catalytic force must then be understood such a mediately exercised force as can be defined according to a general rule through a certain knowledge of the bodies to whose influence the interposing medium is subjected, although without knowledge of this medium itself. The discovered fundamental law of electricity gives a general rule for the determination of catalytic forces in this sense."
* Muncke in Gehler's *Dictionary of Physics*, vol. v. p. 1007.

applied to men of the scientific eminence of Wilhelm Weber or Fechner, particularly from the mouths of literati and *pretended* (*so-genannten*) "men of science," posterity may judge. In the meanwhile we console ourselves with words addressed by Galileo to Kepler:—

"What will'st thou say of the first teachers at the Gymnasium at Padua, who, when I offered it to them, would look neither at the planets nor the moon through the telescope? This sort of men look on philosophy as a book like the Æneid or Odyssy, and believe that truth is to be sought not in the world or nature, but only in 'comparison of texts.' How would'st thou have laughed, when at Pisa the first teacher of the Gymnasium there endeavoured, in the presence of the Grand Duke, to tear away the new planets from heaven with *logical* arguments, like magical exorcisms ! "

Kepler, however, hereupon answered Galileo :—

" Courage ! Galileo, and advance. If I see rightly, few of Europe's eminent mathematicians will fall away from us ; *so great is the power of truth.*"

Chapter Eighth.

THE PHENOMENA SUITABLE FOR SCIENTIFIC RESEARCH—THEIR REPRODUCTION AT DIFFERENT TIMES AND PLACES—DR. FRIESE'S AND PROFESSOR WAGNER'S EXPERIMENTS IN CONFIRMATION OF THE AUTHOR'S.

BEFORE passing on to the description of further experiments and observations which I conducted with Mr. Slade, I may mention that the essential facts (and of these just the most wonderful and incredible) have already been repeated, not in presence of Slade, but among private individuals with medial gifts, under the most stringent conditions. This circumstance disposes first of the argument that Mr. Slade is a swindler and impostor merely on the ground that as a "professional" medium he makes a "business" of his powers like any other conjurer; and secondly, it divests spiritistic phenomena of the exceptional character which might seem to unfit them for becoming objects of scientific research. For the characteristic of natural phenomena is that their existence can be confirmed at different places and times. Thus is proof afforded that there are *general* conditions (no matter whether known or unknown to us, or whether we can provide them or not

* *Wiss. Abh.* vol. iii. (*Transcendentale Physik*), p. 215.

at pleasure) upon which these phenomena depend. It is in the discovery and establishment of these conditions under which natural phenomena occur, that the task of the scientific observer and experimenter consists.

The method applied by me for demonstrating the appearance and disappearance of human limbs by means of sooted paper has proved particularly successful. Paper so treated is like a photographic *camera obscura* which can be placed unobserved and well guarded in the neighbourhood of the medium, so that deception becomes a physical impossibility. In this way Dr. Robert Friese, of Breslau, sitting with a family of that place, a lady with mediumistic powers being present, obtained the impression of a hand upon sooted paper fixed to a slate, which was placed on a stove and covered with a sheet of paper to protect it from dust. The impression was obtained while the medium sat on a sofa between Dr. Friese and a friend of his, and was held by them both. The medium in a state of trance distinctly saw the figure which mounted on the stove and made the impression of the hand, so that the whole operation was described by her during the process, Dr. Friese and his friend perceiving nothing. The slate was taken down from the stove directly the medium awoke, and on it was found the impression of the hand just as she had described it.

But the most brilliant repetition of one of my experiments with sooted slates was achieved in the autumn of last year [*] with a private medium in St. Petersburg. It has been published by Dr. Nicolaus Wagner, Professor of Zoology, and honorary member of the University there, in the June number of *Psychische Studien*, with a photo-lithographic representation of the impression obtained. I reproduce this account here literally, since it also illustrates the ecclesiastical and religious prejudices which now, as in the age of Galileo, attempt to obstruct the work of the scientific investigator.

REPETITION OF ONE OF PROFESSOR ZÖLLNER'S EXPERIMENTS WITH PRIVATE MEDIUMS.

By Nicolaus Wagner, Professor of Zoology, and Honorary Member of the Imperial University at St. Petersburg.

"The reaction against the spiritual movement runs its course with the same violence as every fanatical opposition. If "blind faith" is the motive power of religious fanaticism, so also is the direction of the contrary movement determined by a force which is quite as illogical—"blind scepticism." In the one and the other the cause is the same—feeling, passionately excited, and resisting every cool, matter-of-fact (objective) consideration. There is no better proof of this than the attacks of the *savans* upon those of their colleagues who had the inexcusable temerity to satisfy

[*] 1878.—Tr.

themselves of the reality of mediumistic phenomena, and to publish their experiences to the world. Until their fall into Spiritualism the work and opinions of these men were recognised as entirely logical, accurate, and satisfying the conditions of scientific inquiry. But scarcely have these same scientists carried their researches into the region of mediumistic phenomena, than they are forthwith encountered by the feeling of antipathy ; and that even before the phenomena themselves have been adjudicated upon by sound reason.*

* This reminds us of Mr. Crookes : "It is edifying to compare some of the present criticisms with those that were written twelve months ago. When I first stated in this journal (*Quarterly Journal of Science*) that I was about to investigate the phenomena of so-called Spiritualism, the announcement called forth universal expressions of approval. One said that my 'statements deserved respectful consideration;' another expressed 'profound satisfaction that the subject was about to be investigated by a man so thoroughly qualified as,' &c. ; a third was 'gratified to learn that the matter is now receiving the attention of cool and clear-headed men of recognised position in science ;' a fourth asserted that 'no one could doubt Mr. Crookes' ability to conduct the investigation with rigid philosophical impartiality ;' and a fifth was good enough to tell its readers that ' if men like Mr. Crookes grapple with the subject, taking nothing for granted until it is proved, we shall soon know how much to believe.'

"These remarks, however, were written too hastily. It was taken for granted by the writers that the results of my experiments would be in accordance with their preconceptions. What they really desired was not *the truth*, but an additional witness in favour of their own foregone conclusion. When they found that the facts which that investigation established could not be made to fit those opinions, why,—' so much the worse for the facts,'—they try to creep out of their own confident recommendations by declaring that 'Mr. Home is a clever conjurer, who has duped us all.' 'Mr. Crookes might, with equal propriety, examine the performances of an Indian juggler.' ' Mr. Crookes must get better witnesses before he can be believed.' 'The thing is too absurd to be treated seriously.' ' It is impossible, and therefore can't be.' 'The observers have all been biologised (!), and fancy they saw things occur which never really took place,'" &c., &c.—Crookes's *Researches in the Phenomena of Spiritualism*, p. 22.—*Note by Translator.*

Impelled by this antipathy, even the strongest understanding is blind; it seeks support from and attaches itself to such strangely childish arguments and suppositions, as to any sound thinking and unprejudiced person are in the highest degree absurd.

"In the relations of the *savans* to my colleague, Professor Zöllner, who lately experimented in the mediumistic field, we have the most complete evidence of the justice of the above observation. Satisfied through the force and reality of facts of the entire genuine objectivity of the mediumistic phenomena, he detailed his investigations. But as in the case of the investigations of Crookes and Boutlerow, so were these also forthwith exposed to suspicion, and set down to clever conjuring; and the name of the cautious and accurate investigator swelled the sad list of scientists who had been deceived by (so-called) charlatans.

"Now, since the whole weight of this charge rests on the merely supposed fraud of the mediums, it will not be superfluous if I give to the Press the results of some investigations, analogous to those of Zöllner, which I have made with non-professional mediums. I do not in the least expect that this narrative, any more than hundreds such, will make the slightest impression on the fanaticism of the sceptic: on the other hand, I have the strongest belief that it will serve to confirm the growing conviction of those who are not disinclined to be convinced by the truth of things.

"Since the force of the evidence chiefly depends on the confidence in the mediums, and in the persons composing the circle among whom the *séances* took place, I consider it essential first of all to discuss this question, and to follow it up with some historical statements. Moved by my and my colleague Boutlerow's writings in certain Russian periodicals, the family of the engineer and chemist E——, as also some of their intimate friends and relatives, desired to convince themselves of the reality or otherwise of the mediumistic phenomena. It must further be remarked that in these families earlier cases of a mediumistic character had been already observed, but had been ascribed to different causes, such as accident or hallucination.

"Three ladies took part constantly in the sittings— the wife of the chemist, Sophia E—— ; her sister, A. M—— ; and her friend, A. L——, who had for years been united with Mrs. E—— in the most genuine friendship and sympathy. Of these ladies the two first were gifted with very remarkable mediumistic aptitudes. All three were distinguished by deep religious feelings, and every deception, even for a good end, is abhorred by them as a heavy sin. The manifestations occurring almost from the very first were regarded by them as miraculous, and this feeling was confirmed as the phenomena became more and more developed.

"The fourth lady, who was likewise constantly present, was Miss Catherine L——; one of the greatest friends of Sophia E—— the wife of the chemist E——. At the commencement of the *séances* she was an atheist; all her convictions leaned to materialism. She held the principles of the well-known Russian publicist, Herr Pisaref, as irrefragable dogmas. The power of the manifestations shook, and at length overthrew, this fanaticism of hers.

"This small circle was formed with the firm expectation that it would succeed in demonstrating the mediumistic manifestations to be simply a further development of already-known physical phenomena. With this object the table at which they sat was placed upon glass supports, and round the feet of the table was wound a wire, the ends of which were attached to a galvanometer. Instead, however, of the expected physical phenomena, the table at the very first *séance* urgently demanded the alphabet, and by means of blows with the foot of the table the following sentence was spelled out:—

"'I suffer because thou believest not.'

"'To whom does that refer?' asked those present.

"'To Catherine L——.'

"'Who, then, art thou?' asked L——.

"'I am thy friend, Olga N——.'

"This dearly-loved friend, also an atheist, had died about a year before, and on this account Catherine L—— was deeply astonished and moved by the information communicated through the table. This information, given in the same *séance*, referred to different particulars of an event known only to Catherine L——, and thoroughly convinced her of the existence of the soul of her beloved friend, even though in another world.

"Henceforth, the before-mentioned physical experimentation was laid aside, the conversations were more and more striking, and confirmed their faith in the reality of another world. This faith soon became a firm conviction with all. To show the relations of the circle, and especially of Catherine L——, to the phenomena, I here add some extracts from her diary, which was written for her own eye only, and communicated to me after her death, which happened somewhat later.

"'29th March, 1876, 1.30 A.M. Scarcely had S—— and I retired to rest, and left off talking that we might sleep, than suddenly there sounded a beating on the wall at the head of my bed. I supposed at first that some one was probably passing on the stairs adjoining my wall, but after some minutes the knocking was repeated, and with such force that S—— also became attentive, and asked me if I had knocked. Now I guessed what it was.

"'Probably my Olga is now come to me,' said I. In assent sounded immediately three times, one after the other, a muffled blow, as if a soft wall had been struck with a hammer wrapped up in something soft.

"'Is it thou, Olgchen?' I asked the spirit aloud. Three regular knocks answered.

"'Can I sleep quietly this night?' Again the like three knocks.

"30th March, 6. 45 P.M.

"'Why did you knock at my wall yesterday, Olinka?'

"'Evil spirits prevent you going to the supper. Thou wouldst do it, and hast abandoned this intention. I came yesterday to say to thee that thou, dear one, shouldst not obey them. I will not come for a whole week. I have much to do. On Thursday, after the supper, I will visit thee.'

"'So, if I take the supper, thou will'st come to me?'

"'Yes! and I will make thee a present.'

"'What sort of a present?'

"'Thou canst show it to every one.'

"'Thou will'st give it to me on the day of the Communion?'

"'Yes, in the church.'

"First of April. I have confessed. After the supper I went and took my place in the church. Suddenly in my hand there came a nosegay of white rose and myrtle, tied with a lock of the dear and well-known hair! That was the promised present.

"'Come home from church, we sat ourselves at the table. Our heavenly friend was already among us. Her first words were—

"'I wish you all happiness. I am happy for you. My darling! art thou content with my present?'

"'What significations have the rose and myrtle?'

"'Pure love. Eternity.'

"'I could scarcely restrain my tears.

"'30th April, 10 o'clock. S. E——, sitting on a chair, fell into a trance, of which the spirit informed us. Afterwards a hand was shown to us, one after the other; at our wish it touched our hands, and came close to the sight of those of us who had not been able to distinguish it clearly enough. I asked the spirit whether I could kiss this hand? The spirit replied that its hand would be between the table and the cloth, and that I might kiss it through the cloth. Twice I kissed the dear hand, and convinced myself thereby thoroughly of its reality: it was a living, flexible hand.'

"I have given these extracts to show the genuine and cordial relations of the deceased to these observers of the phenomena which took place before their eyes. Again, I repeat, that she wrote her diary for herself alone, and probably never thought of the possibility that extracts from it might appear in the Press. The circle itself, in the sittings of which she took part, was exclusively interested in the phenomena for their own sake, and was utterly and altogether unconcerned with the spiritualistic propaganda. All the usual mediumistic phenomena, such as the self-moving of objects, lights, appearance of hands, &c., took place at these *séances*. Especially often were objects brought to the circle, most frequently pictures of saints, hair, and flowers. During a *séance* in the spring the whole table was literally covered with flowers. During another *séance* the daughter of Sophia E——, a young lady of fourteen years old, received a live green frog, to make up for the loss of one that had died a few days before. This frog remained with her for some days alive, and afterwards disappeared.

"On one occasion the spirit of Olga N—— declared that she would fully materialise, and designated Sophia E—— as the strongest medium, through whose means the materialisation would be effected. On the evening appointed by the spirit the medium was laid upon a sofa and separated from the rest by a curtain formed by hanging up a plaid. She remained, however, so far visible that her position could always be observed. It was half dark in the room. After the medium had fallen into a trance she was several times raised in the air, placed upon the boards, and again carried back to the sofa. Afterwards a white figure, covered with a thick veil, was raised behind and above the curtain. Quietly, calmly, it came over the curtain to the table at which the party were sitting. Then it went to Catherine L——, embraced and kissed her, touched her face with its hand, and disappeared, whilst raised again in the air. At the next *séance*, which was in darkness, the phenomenon was repeated, and Catherine L—— was covered with a veil, which was left behind upon her.* After this phenomenon the sittings of the circle almost ceased. Amazed with what they had seen, they were all convinced that it would be a sin, after these proofs of the reality of another world and of a higher power, to continue the *séances*, though at the same time they did not refuse individual communications and instructions from that world, and for this purpose availed themselves from time to time of the usual means of intercourse, such as table tipping and psychography. Of course, therefore, the phenomena did not cease, and they were not seldom concerned in different events which happened to the families of the mediums.

"All this had gone on for about a year, up to the winter of 1877, when I accidentally made the acquaint-

* The condition of the medium during the trance made a deep impression on all present, the most lasting one, naturally, upon her husband. After the *séance*, she was for some days ill; at the same time there appeared upon her left side a broad blood-swollen spot. (Compare the description of a materialisation from the left side of the medium, Dr. Monck, in *Psychische Studien*, 1877.—Note by the Editor). These unfortunate results were supposed by the sitters to be owing to their having put forward the *séance* earlier than the appointed time.

ance of the chemist E—— and his family. Entertaining the wish to receive some proofs of the objectivity and reality of the phenomena, I begged some of those who had taken part in the earlier *séances* to afford me the opportunity. I obtained their entire consent, and found the greatest readiness to comply with my wish, although the sentiments of the whole circle were openly opposed to my opinions. This opposition was especially marked in the case of Catherine L——, who, as compensation for her discarded materialism, was now fanatically addicted to ultra-orthodoxy. She continually maintained against me that no evidences of these things could ever convince any one, since they were matters of faith and not of knowledge. Such being the relations of the circle, it was not to be expected that we should obtain any decided results.

"During the first sitting in which I took part, and which was held in a dim light, a hand was formed above the small table, which was covered with a cloth, and afterwards came out from under the cloth, remaining above the table some minutes, and, gently moving, touched those who inclined themselves towards it. This was the only materialisation, and the only remarkable phenomenon in the series of not very numerous *séances* which lasted up to the end of the winter.

"Catherine L—— had long suffered from a chronic

catarrh, which at this time took the form of consumption. Her disposition was still hostile to my objects, so that we were compelled to give up the *séances*. She died in the arms of Mrs. Sophia E——, amidst the proofs of her love, friendship, and affection.

"In the autumn of the year 1878 the relations of the circle to the mediumistic phenomena were completely changed. After the spirit of the deceased Catherine L—— had given consent to the continuance of the sittings, and promised good success, remarking, however, at the same time, that the results would be received with distrust, the circle was widened by the addition of some young people; the engineer, the mechanist M——, was one of the constant sitters; sometimes the physician L—— took part in the sittings.

"In the very first sitting we were directed by raps to repeat the experiment of Professor Zöllner; and since it is the object of this publication to confirm that experiment, I will not dwell upon other more or less remarkable phenomena which occurred at our *séances*.

"We took an ordinary folding slate, with clasps; on each side within was fastened, by means of wax, paper, blackened with soot. The slate was then tied together with a string, and the ends of the string, as well as the edges of the slate, were fastened with four seals with the signet of the chemist E——, and the signet was entrusted to me for safe custody at home. We were informed by means of raps that this slate must lie upon the table for four *séances*, and impressions would then be found upon it. At the *séances* the table was always covered with a cloth, and between this and the table the slate was laid. With the development of the phenomena the slate began to move of itself. It went from one to the other, in order that it might remain for some minutes under the hands of each of those present.

"In the third sitting we were enjoined to seal the slate with seven seals, with another signet of the chemist E——. We asked, 'Is there anything on the slate?' It was answered, 'I do not know.' Thereupon we asked if we might open it? The answer was, 'Yes, you can.' We opened the slate; both papers were untouched. We closed it again, bound, and sealed it with seven seals. The signet I again took away with me. At the following sitting violent movements of the slate again occurred, and finally I was directed to lay the slate on my knees. I did so, and then placed my hands again upon the table. For some minutes the slate remained quiet; then I had the sensation as if some one lightly touched it for a while. Soon after we were told, through sharp decided raps, to take away the slate. To the question, 'Is there anything on it?' a strong definite affirmative answer was returned. 'Can we open it?' 'Yes!'

"We struck a light (the *séance* was in the dark), opened the slate, and perceived an impression on each side: upon the right, that of a hand; upon the left, that of a foot. All three female mediums and the chemist E—— at once recognised in the impression the hand of Catherine L——, which had characteristic peculiarities. It was unusually large and long for a female hand, the little finger being strongly bent out. The foot, also unusually large, could not find room enough on the slate, and this impression, moreover, was not very clear. The hand was much more sharply impressed, if not quite so distinct, as was the case with Zöllner's impression. (I here add the copy of our impression.*) For greater certainty this impression was shown to a sculptor, who well knew the hand of the deceased, and he at once asked, 'Is this an impression of the hand of Catherine L——?' He supposed that the impression had been taken during her life. We were apparently ourselves partly to blame for the want of distinctness in the impression. Every one who is familiar with mediumistic phenomena knows their whimsicality, and that promises given at *séances* are often not fulfilled. Not much expecting success, we went only superficially to work in the preparation of the slate—did not fasten the paper with smooth

---

* There is here a foot-note relating to the illustration omitted in this translation, the illustration itself, as two or three others, not seeming indispensable.—*Tr.*

regularity over the sides of the slate, and the soot was not thickly and regularly spread. Had we only found anything when we opened the slate after the third *séance*, which would have given us but a remote assurance of future results, we should then have rectified every defect in our preparations.

"The above-recorded objective proof I regard as sufficient to obviate every suspicion of deceit. Had it been even possible to imitate the seal and open the slate, yet was it at any rate impossible, and indeed without aim or object, to imitate the impression of the hand of the deceased C. L——. All those who took part in the *séance* were "believers;" all were in like manner interested in the experiment: no one among them was so depraved and mischievous as to contrive so cruel a mystification, cruel in relation to the persons to whom the memory of the deceased was sacred. That young lady was more than a relative in the family of the chemist E——; one could not but see the joyful rapture of the mediums at the moment when they recognised in the impression the hand of Catherine L——. All crossed themselves and wept: all regarded this result as a miracle.

"After this phenomenon some of those present proposed to terminate the sittings, since no better, more objective, more all-convincing, complete proof could be obtained; I, however, wished to continue them, though they must at all events soon cease.

The next sitting had already lost the characteristics of our usual *séances*. The phenomena were languid and intermittent. The spirit of Catherine L—— declared that it could not appear for a whole month. Other disturbing circumstances concurred, so that we resolved to postpone our *séances* to a more opportune time. An unexpected misfortune intervening compels us to renounce them for a long time, perhaps for always.

"In giving this simple history, with its childlike full conviction and faith in the personality of the spirit (Fetishism), I repeat that it can have no effect upon the stubborn scepticism of those who have become the slaves of their *à priori* convictions. This narrative can have only an irritating tendency with such, excite their scepticism up to a fanatical point, and drive them, even should they admit the facts, to discover some explanation even more senseless than Carpenter's 'unconscious cerebration.' But those with whom Fetishism * is no subjective product of our brain and feeling, who recognise the necessity and legitimacy of individuality as the lever of the development of humanity and of well-being, those will find in these facts the proof and confirmation of their views.

* "Fetischismus" is the word used, but not, it is conceived, in the sense that word bears in English.—*Note by Translator*.

"Again, these facts convince us of the necessity of widening the domain of recognised science and its methods and means for the exploration of the invisible and unknown world, of the existence of which we have in our hearts from childhood so clear, so simple, and so warm a presentiment.   N. WAGNER."

Chapter Ninth.

THEORETICAL; "THE FOURTH DIMENSION"—PROFESSOR HARE'S EXPERIMENTS—
FURTHER EXPERIMENTS OF THE AUTHOR WITH SLADE—COINS TRANSFERRED
FROM CLOSED AND FASTENED BOXES—CLAIRVOYANCE.

PASSING on to the account of further experiments with Mr. Slade, I take those in the first instance which I had devised on the principle of the extended conception of space* (*Raum-anschauung*) — for the purpose of experimental proofs of the reality of a fourth dimension.

Among these proofs there is none so instructive and convincing as the transport of material bodies from a space enclosed on every side. Although for our three-dimensional intuition such a space apparently allows of no other exit than through the material boundaries, yet from the fourth dimension it can be opened, and thus the transport of the body in

---

* The philosophical sense of the word intuition (*Anschauung*) may have some difficulty for non-metaphysical English readers. With us it usually denotes an internal sense; in German philosophy it is the act of perception, whether of the external or internal sense, before all application of the categories of the understanding by which the "matter" of the perception becomes an "object." In the Kantian philosophy, which Zöllner follows, space and time are intuitional "forms." The word *Anschauung* is translated intuition and conception at different places as the context seems to require.—*Tr.*

this direction can be effected without disturbance of the three-dimensional material walls. Since the so-called intuition of a four-dimensional space is wanting to us, as three-dimensional beings, we can only form to ourselves a conception of this proceeding by an analogy taken from the next lower region of space. Suppose in a plane a figure of two dimensions enclosed by a line on every side, in which is a movable object. By movements *only* in the plane that object could not escape from the interior of that *two-dimensionally* enclosed space otherwise than by an opening of the line of enclosure. But if the object were capable of a movement in the third dimension, it would need only to be raised perpendicularly to the plane, to be passed over, and let down again on the other side of the line. To two-dimensional beings who reasoned on the assumption that *only* such movements were possible as they could *intuitively represent* to themselves, *i.e.*, only two-dimensional movements, the proceeding just described would seem a miracle. For the body which they suppose to be *completely* enclosed must at a certain spot transiently *vanish* for them, in order suddenly to reappear at another spot. Although similar facts have been so frequently observed at spiritualistic *séances*, and publicly testified to by the most credible and intelligent men, yet as an introduction to the description of my own experiments, I cannot omit to impart the following fact observed by the celebrated American scientist and chemist, Professor Hare.*

It is that described by State Councillor Aksakow in *Psychische Studien* (edited by him) in the July number of this year (1879), under the title "Some Experiments of Professor Hare Confirmatory of Zöllner's Experiments." I confine myself here to the first experiment, described in a letter published on the 1st May 1858 by an eye-witness, Dr. S. A. Peters, who had visited Professor Hare in his laboratory, in order himself to witness some of the remarkable phenomena which Hare had publicly reported. The letter

---

* Robert Hare, Doctor of Medicine, Professor of Chemistry at the University of Pennsylvania in Philadelphia; born 1781, died 15th May 1858. In Poggendorff's *Literary Biographical Dictionary*, from which I have taken the above particulars, will be found a catalogue, filling a whole column, of Hare's numerous chemical and physical treatises. In text-books of Physics his name survives in the so-called "Hare's Spiral," a galvanic element, in which a copper and zinc plate, properly separated by bad conductors, are rolled over one another for the production of the greatest possible surface. With this arrangement, previously to the construction of constant batteries, very strong effects of light and heat could be produced. The treatise of Hare's referred to is published in Tilloch's *Philosophical Magazine*, of the year 1837, under the title "New Voltaic Battery."

In his later years Professor Hare undertook, as a *true* man of science, the most thorough experimental investigation of the phenomena of Spiritualism, for which, in his country, convenient opportunities offered. He even evinced his acuteness in this field in the construction of suitable apparatus and instruments. One of these he named the "Spiritoscope." It consisted of an apparatus connected with a cipher-plate and index, similar to that which was applied to the first electric telegraphs. A detailed description, with picture, of this ingenious apparatus, in which the motions of the index are completely concealed from the medium, will be found in the pamphlet, *Experimental Investigations of Spirit-Manifestations*, by Dr. Robert Hare, Professor of Chemistry, &c., &c." German edition by Alex. Aksakow, Leipzig, 1874, Mutze.

is addressed to the editor of *The Spiritual Telegraph*, and is as follows :—

"PHILADELPHIA, *April* 18, 1858.

"Mr. Editor,—Finding myself in this city on a visit from the State of Missouri, I availed myself of the opportunity to visit Professor Hare, in order to see what new developments or discoveries he has made in Spiritualism. I have no doubt that a history of the most astonishing spiritual manifestations which are now taking place in the Professor's laboratory will shortly be given to the public.

"I will now confirm what I saw myself. Dr. Hare, the medium (a young man named Ruggles, of from eighteen to nineteen years, to whom I was quite a stranger when I entered the laboratory), and myself, were the only persons present. The medium sat down in front of the *spiritoscope*, which stood on the table in the middle of the room. Dr. Hare and I sat opposite and close to the table. After some minutes it was said to us through the spiritoscope, 'Let Dr. S. A. Peters put two glass tubes and two pieces of Russian metal in the box.' Dr. Hare thereupon left his seat, and fetched me two glass tubes of about six inches length, and half an inch diameter, hermetically sealed at the ends, and also two pieces of Russian platinum, each of the shape of a common musket-ball. I first examined the box in which I was to deposit these objects. It stood on the table before me. It resembled a writing-desk; was about two feet long and half a foot broad, four to eight inches deep, and had a lid which let down slantwise, with hinges and a lock. In this box I placed the two glass tubes and the balls of platinum—there was nothing else in it,—and locked it. Dr. Hare and I then took our seats as before, and the medium, Mr. Ruggles, continued at the spiritoscope. After the lapse of fifty-five minutes there was said through the spiritoscope, 'We have a present for Dr. S. A. Peters; let him go to the box and fetch it.' Hereupon I went to the box, which was only a single foot from me, opened it, and found—*the two pieces of Russian platinum inside the two hermetically sealed glass tubes.*

"I will make no observations on the above. What I have seen I hold it to be my duty to make known to the world. I have no other interest in making the above statement than the desire to serve my fellow-men. S. A. PETERS."

I go now to the account of similar experiments, which have succeeded with me in the presence of Mr. Slade, but which have a yet higher interest for me, in that they have produced in me a conviction of the reality of the so-called clairvoyance, or clear-seeing.

On the 5th May, 1878, at about twenty-five minutes past four, Mr. Slade, Herr Oscar von Hoffmann, and

I, took our places at the table and in the sun-lighted room, of which a photographic copy is seen in the frontispiece. Besides a number of slates, purchased by myself, there lay upon the table other things, among them two small cardboard boxes, in which, at Slade's first residence in Leipsic, in December 1877, I had put some pieces of money, and then firmly plastered it up outside with strips of paper. I had already at that time been in hopes of the removal of the enclosed pieces of money without opening of the boxes. However, my friends and I were so astonished and occupied with the multitude of the other phenomena which happened at Slade's first and second visits to Leipsic (November and December 1877), that I abandoned the above-mentioned experiment for the time, and postponed it till Slade's return to Leipsic. One of these boxes was in form circular, and within it was a large piece of money; this box was firmly fastened by a strip of paper, the breadth of which corresponded to the height of the box, and its length much exceeded the circuit of the box; so, indeed, that first the strip of paper was spread with liquid glue on one side over its whole length and breadth, and was then stuck several times round the box, so that the latter, after the fastening, presented the appearance of a low cylinder of pasteboard. The other box was rectangular, of the same sort as those in which steel pens are kept. In this box I had put two small pieces of money, and had then closed it by sticking a strip of paper round it, perpendicularly to its length, by means of liquid glue.

As mentioned above, I had already, in December 1877, fastened up these boxes, and as I had observed neither the value of the enclosed coins nor their date, I could afterwards only ascertain by the noise from shaking the boxes, that enclosed in the circular one was a large German coin (a thaler or a five-mark piece), in the rectangular one two smaller coins; whether these were pennies, groschen, or five-groschen pieces I had, after the lapse of half a year, at the time of Slade's last stay in Leipsic, entirely forgotten.

After we had taken our places at the card-table on the above-mentioned day in the manner described, I took up the round box, and satisfied myself, by shaking, of the presence of the coin I had enclosed in it. Herr O. von Hoffmann did the same, and lastly Mr. Slade, who asked us for what purpose I had designed this box. I explained my purpose in a few words, and at the same time declared that it would be one of the finest confirmations of the reality of the fourth dimension, if his invisible, intelligent beings succeeded in removing that coin from the box without opening it. Slade, ready, as always, to conform to my wish, took in the usual manner one of the slates which lay at hand, laid a morsel of slate-pencil upon it (indeed, as it happened, a considerably larger one than usual), and held the slate with his right hand half under the table. We heard writing, and when the slate was drawn out, there was found upon it the request to lay a second piece of pencil on the slate, which was done. Then Slade, who sat at my left (Von Hoffmann was on my right), held the slate with the two bits of pencil again under the table, while he as well as we awaited intently what should come there. Meanwhile the two fastened-up boxes lay untouched on about the middle of the table. Some minutes passed without anything happening, when Slade gazed fixedly in a particular direction in the corner of the room, and at the same time said, quite astonished, but slowly, the words dragged after one another, and partly with repetition: "I see — see funf and eighteen hundred seventy-six." Neither Slade nor we knew what that could mean, and both Herr O. von Hoffmann and myself remarked almost simultaneously that, at any rate, "funf" signified "funf" (five), and made the sum of the addition $5 + 1876 = 1881$. While I threw out this remark half in jest, we heard a hard object fall on the slate, which Slade during all the time had held under the table with his right hand (the left lying before us on the table). The slate was immediately drawn out, and on it was found the five-mark piece, with the date 1876. Naturally I forthwith snatched up the pasteboard box, standing before me, and which during all the foregoing had been touched by nobody, to ascertain, by shaking, the absence of the piece of money which had been in it half an hour before; and behold! it was quite empty and silent; the box was robbed of its contents in the shape of the five-mark piece.

As may be supposed, our pleasure at such an unhoped-for success of our experiment was extremely great; all the more, that by it at the same time was established the existence of a direct perception of objects, not effected in the ordinary way of our sense-perceptions.

Moreover, it could not be any so-called thought-reading by the medium; that is, the perception of representations already in the heads of human beings. For neither I, and much less Mr. Slade and Herr von Hoffmann, knew what sort of coin there was in the box, nor what date it bore.

I was so satisfied with the success of this experiment under such stringent conditions that I was thinking of putting an end to the sitting, and postponing further attempts to a later one. However, Slade remarked that he did not feel himself at all exhausted by the sitting, which had lasted at most ten minutes. This remark of Slade caused us to keep our places at the card-table, and to engage in unconstrained conversation with him. I introduced the subject of his sitting with the Grand Duke Constantine of Russia, and requested him to give us a detailed account of the phenomena which took place at it, as hitherto we had seen only the brief paragraph statement about them in the press. Thus urged, Slade mentioned that a very remarkable experiment in slate-writing had succeeded in the presence of the Grand Duke Constantine. Accidentally there had been two bits of pencil on the slate ; when he held it under the table the writing of two pencils was heard at the same time, and when he drew out the slate the one pencil had written from left to right, the other, *at the same time*, from right to left. I at once proposed to try whether this experiment would succeed also with us : the suggestion arose from me quite naturally, from the association of ideas elicited by the two bits of pencil which had been required in the above-mentioned experiment, without our having as yet known the object of this written demand.

Slade, at once ready to comply with my wish, held the slate with the two bits of pencil under the table-surface, and we soon heard, very clearly, writing upon it.

When the slate was withdrawn there was on it a communication in English as follows :—

"10—Pfennig—1876
2—Pfennig—1875.

Let this be proof to you of clairvoyance. After the nine days you must rest, or it will harm you and the medium. Believe in me, your friend."

P. 160-61.

We at once referred the first part of this message to the two coins contained in the rectangular box still unopened. I was just about to open it, we having immediately before convinced ourselves by shaking the box and the distinct jingling within it, of the presence of the two smaller coins, yet without knowing the value or date of them. Suddenly, however, I changed my intention, and set the little box again uninjured on the middle of the table, while as well Herr von Hoffmann as also Slade suggested the possibility that perhaps the two coins, in like manner as shortly before the five-mark piece, might fall from the unopened box upon the slate held underneath. Immediately upon this suggestion, Slade again held an empty slate under the middle of the table. Scarcely was this done, when we distinctly heard two coins drop down on the surface of the slate, and on closer examination, the above statements on the slate we, in fact, found confirmed. Highly delighted, I now seized the still closed box in the confident expectation that it would, like the round box, be empty, and that, therefore, on shaking no rattling within would be heard. How great was my surprise when nevertheless the rattling happened, proceeding, indeed, likewise from two bodies, which yet, judging from the altered character of the sound, could not be coins. Already I was intending to convince myself of the contents of the box by opening it, which could not be done without tearing the strips of paper pasted over it, when Slade prepared to get our question answered, as usual in such cases, through slate-writing, by his "spirits." Scarcely had he taken a slate with a fragment of pencil lying upon it, and held it half under the table, when we distinctly heard writing. Upon the upper surface of the slate was written in English—

"The two slate-pencils are in the box."

In fact the two larger pieces of slate-pencil were nowhere to be found, and when I now opened the box by tearing the strip of paper glued to it, there within it, to our delight, were both the pieces of pencil.

The foregoing facts are of great value in a threefold aspect. *First*, there is proved the occurrence of writing under the influence of Slade, the purport of which was necessarily *unknown* to him before. It is consequently *impossible* that these writings occur under the influence of the *conscious* will of Slade, whatever *modus operandi* is presupposed.

Secondly, the apparent, so-called passage through matter is proved in a highly elegant and compendious manner. In order to reach by the shortest way the surface of the slate, the coins must apparently have penetrated not only the walls of the box, but also about 20 millimètres thickness of the oak table. The two slate-pencils must have travelled the same way in a *reverse* direction from the surface of the slate.

Thirdly, by these experiments an incontrovertible

proof is afforded of the reality of so-called *clairvoyance*, and that in a double way. The first time, with the five-mark piece, the contents of the closed box appeared in the form of a definite represented image in Slade's intuitional life; he "saw" the numbers 5 and 1876. The second time this was not the case; but the contents were communicated to us in the form of written characters on the slate. The contents of this rectangular box must therefore have existed as imaged in another, not a three-dimensionally incorporated intelligence, before that represented image could be transmitted to us by the aid of writing. Hereby is proved, as it seems to me, in a very cogent manner the existence of intelligent beings, invisible to us, and of their active participation in our experiments.

I have already shown that the whole phenomenon of clairvoyance admits of a very easy and natural explanation by help of the fourth dimension. From the direction of the fourth dimension, the, to us, three-dimensionally enclosed space must be regarded as appearing open, and indeed in an interval from the place of our body so much the greater, the higher the soul is raised to the fourth dimension. At the same time, with the increasing elevation to this fourth dimension there is a widening of the overlooked space of three dimensions, just as by elevation above the surface of the earth there is, according to geometrical laws, a widening of the overlooked two-dimensional expanse. Also in three-dimensional space, representations of the changes of place of our body at rest, as for example, when we are in the car of a balloon, are produced merely by changes in the appearances of objects. The manner in which this happens, for instance, in our present organization by help of the sense of sight, is only a modification of the above-mentioned general fact, and depends, as observed, upon the changeable forms of the organic disposition; that is, of our body, through which the representations of our soul are mediated. Thus, Slade's soul was, in the first case, so far raised in the fourth dimension that the contents of the box in front of him were visible in particular detail. In the second case, one of those intelligent beings of the fourth dimension looked down upon us from such a height that the contents of the rectangular box were visible to him, and he could describe its contents upon the slate by means of the pencil.

It is of interest to compare the theory of clear-seeing here indicated, with the description of this condition by persons who have become clairvoyant by being thrown into the so-called magnetic sleep by a "magnetiser." In accordance with the above theory, and the principle of continuity, we should expect that from the beginning of the clairvoyant condition its increasing development must be attended with a spatial widening of the three-dimensional circle of sight, that is, bodies must gradually become transparent in continually greater intervals; quite in analogy with the increasing number of objects which we perceive by continual elevation above the earth. This supposition appears to be confirmed in fact by the descriptions of the American clairvoyant Davis, who depicts his perceptions in the magnetic sleep and otherwise in the following words : *—

"The sphere of my vision now began to widen. . . . Next, I could distinctly perceive the walls of the house. At first they seemed very dark and opaque; but soon became brighter, and then *transparent;* and presently I could see the walls of the adjoining dwelling. These also immediately became light, and vanished—melting like clouds before my advancing vision. I could now see the objects, the furniture, and persons, in the adjoining house, as easily as those in the room where I was situated. At this moment I heard the voice of the operator. He inquired, 'if I could hear him speak plainly :' I replied in the affirmative. He then asked concerning my feelings, and 'whether I could discern anything.' On my replying affirmatively, he desired me to convince some persons who were present, by reading the title of a book, with the lids closed, behind four or five other books. Having

* *The Magic Staff*, an autobiography, by Andrew Jackson Davis, p. 217 (13th edition), New York, 1876.

tightly secured my bodily eyes with handkerchiefs, he then placed the books in a horizontal line with my forehead, and I saw and read the title without the slightest hesitation. This test and many experiments of the kind were tried and repeated; and the demonstration of vision, independent of the physical organs of sense, was clear and unquestionable. . . . But my perceptions still flowed on! The broad surface of the earth, for many hundred miles, before the sweep of my vision—describing nearly a semicircle—became transparent as the purest water; . . . and I saw the brains, the viscera, and the complete anatomy of animals that were (at that moment) sleeping or prowling about in the forests of the Eastern Hemisphere, *hundreds and even thousands* of miles from the room in which I was making these observations."

Confidence in the above description of subjective representations in the magnetic and clairvoyant condition being presupposed, and that these representations can also be repeated and confirmed, *according to a law*, by other individuals under other conditions in the clairvoyant state, there would be connected with the increased duration and depth of the magnetic sleep an extension of the field of our cubical vision similar to the extension of our field of quadratic vision, according to the laws of perspective, which is associated with elevation over the earth.

The ascertainment of these laws of perspective for

space intuition widened by a dimension would first of all be the task of *geometry*, just as the elements of Euclid must have been known and have become the common property of physicists and astronomers before the spatial significance of celestial phenomena could be thought. That intuitional images, or representations of objects of sight clothed with all the attributes of sense, arise, change and disappear in our soul without the intervention of the physical sight is proved by dreams, hallucinations, and illusions. Of the causes by which these representations are produced in us we know nothing, and can therefore only advance hypotheses about them. But if we ask ourselves wherein consists the difference between these images and those which are produced in us in daily life by means of the sense of sight, we see that there is greater vivacity, regularity, and continuity in the latter. But the essential criterion for the fact that the latter class of representations have real objects in an external world corresponding to them, is the *geometrical* criterion; that is, the possibility for our understanding to refer a part of the changes and differences of those representations to the geometrical laws of remoteness and position. Should the same hold good of the former class of representations—those not obtained through the sense of sight—we should be compelled to relate these also to real objects in an external world, no matter whether the requisite geometrical laws are to be sought in space as *hitherto* conceived, or as widened by a dimension.

In both cases, however, the *causes* by which those images are produced remain unknown to us so long as their *homogeneity* cannot be experimentally proved. We know from internal experience that our will is able up to a certain degree, by means of the so-called force of imagination, to produce at pleasure representations of objects of sight in *our own* soul. In this case we recognise our *own* will as the cause of our representations. If, now, experiments could be instituted, in which this individual will of a single man could produce in like manner, at pleasure, representative images in the soul of another, spatially separate from the willing subject, these images being clothed with all the attributes of reality which we ascribe to the so-called real or actual world surrounding us, thereby would experimental proof be afforded that the phenomenon of a *real external world* can be produced and evoked by an *individual* will, matched with *intelligence*, in another individual. But in that case it would be a *necessary* conclusion, according to the principles of scientific induction, to accept also a qualitatively similar cause for the representation of our whole *real corporeal world:* that is, an *individual* will combined with *intelligence*, however much that individual will may excel the human will in strength and intelligence *quantitively*. I maintain that the above induction is scientifically and logically necessary, and, moreover, the only possible one open to a rationally operating understanding. Newton asserts the same in the third book of his Principia in the third *Regula Philosophandi* in the following words:

"*Ideoque effectuum naturalium ejusdem generis eædem assignandæ sunt causæ quatenus fieri potest; —utique respirationis in homine et in bestia; descensus lapidum in Europa et America; lucis in igne culinari et in Sole; reflexionis lucis in terrâ et in planetis.*"

"Therefore, the same causes are to be assigned, as far as possible, of natural effects of the same kind;— as of respiration in man and in beast, of the descent of stones in Europe and in America; of light in a kitchen fire as in the sun; of the reflection of light on the earth and in the planets."

In the above case there remains only the question whether it is experimentally demonstrable that the human will is able to induce such vivid representations in the consciousness of another, that the latter regards them altogether as he regards the representations whose causes we ordinarily designate as *real objects*, or "bodies." Experiments of this kind have been, in fact, publicly instituted in Germany, by the magnetiser Hansen, of such a surprising and convincing nature, that it is impossible to doubt the reality of this influence of an individual intelligent will upon another, spatially distinct, individual.* Consequently our understanding is constrained, according to the laws of scientific induction and Newton's third *Regula Philosophandi*, to accept an individual will, joined with intelligence, as cause and author of that world of representations which surrounds us in daily life, the so-called *real* external world, or Nature. Whether this intelligent will in the production of our human world of representations makes use of numerous other individual and intelligent existences, or whether there is no other individualisation of intelligent will in nature than human beings and animals, so that that highest author of our real represented world alone influences us, *directly* and according to harmonious laws, is for the present a question of secondary importance. This only is established, that an *individual* being, gifted with intelligence and will, must be presupposed as the cause of our real world of representations.

I may here point out that the foregoing inductions are not new and peculiar to myself. The priority is incontestably due to the English philosopher Berkeley, a contemporary and follower of Newton. In his celebrated treatise "On the Prin-

* Later, in the same volume, the author gives a detailed account of the experiments referred to. The English reader will find abundant evidence on the subject in the late Professor Gregory's book on Animal Magnetism, of which a new edition has been recently published. Harrison; London, 1877.—Tr.

ciples of Human Knowledge" (section 33), Berkeley remarks :—

"The ideas impressed on the senses by the author of nature are called *real things:* and those excited in the imagination being less regular, vivid, and constant, are more properly termed *ideas*, or *images of things*, which they copy and represent. But then our sensations, be they never so vivid and distinct, are nevertheless ideas, that is, they exist in the mind, or are perceived by it, as truly as the ideas of its own framing. The ideas of sense are allowed to have more reality in them, that is, to be more strong, orderly, and coherent, than the creatures of the mind; but this is no argument that they exist without the mind. They are also less dependent on the spirit, or thinking substance which perceives them, in that they are excited by the will of another and more powerful spirit; yet still they are *ideas*, and certainly no idea, whether faint or strong, can exist otherwise than in a mind perceiving it." Corresponding to this deduction, Berkeley, in his 30th section, has the following, on the signification of the laws of nature: "The ideas of sense are more strong, lively, and distinct, than those of the imagination; they have likewise a steadiness, order, and coherence, and are not excited at random, as those which are the effects of human wills often are, but in a regular train or series—the admirable connection whereof sufficiently testifies the wisdom and benevolence of its author. *Now the set rules, or established methods wherein the mind we depend on excites in us the ideas of sense, are called the laws of nature; and these we learn by experience, which teaches us that such and such ideas are attended with such and such other ideas, in the ordinary course of things."*

After this digression, I return to my experiments with Slade, and will first describe another experiment, by which the above facts, with some modifications, were essentially confirmed.

Chapter Tenth.

AN EXPERIMENT FOR SCEPTICS—A WAGER—SLADE'S SCRUPLES—A REBUKE BY THE SPIRITS—AN UNEXPECTED RESULT—CAPTIOUS OBJECTIONS.

In order to satisfy others, who have not personally taken part in the sittings with Slade, of the reality of the phenomena, especially of the slate-writing within a locked double, or 'book,' slate, I hit upon the following expedient. I had bought at the paper and office-utensil warehouse of F. G. Mulius, in this place (Market No. 13), a great number of such book-slates provided with hinges. These bear inside on the polished wooden frames the manufacture mark " A. W. Faber, No. 58," are rectangular, and their outer extent amounts to 260 millimètres in length and 184 millimètres in breadth. Since the breadth of the wooden frames is 20 millimètres, there remains for the size of the two interior slate surfaces a rectangular surface of 220 millimètres in length, and 144 millimètres in breadth. Since the plane of the wooden frame overtops that of the slate-surface within, on each side, by about 3 millimètres, there is within such a book-slate, when completely closed, a free space of 220 millimètres length, 144 millimètres breadth, and 6 millimètres depth. At the side where the hinges are (which are very solid, of brass, and 20 millimètres broad), the edges of the wooden frame shut together so tight that it is impossible to pass between them any object of appreciable thickness (for example a single sheet of writing-paper), and so to introduce it into the inner space of the closed slate. Moreover, the interval between the brass hinges—each fastened by six wooden screws—is only 122 millimètres. On the front side, each of the two wooden frames has a pierced cylindrical brass spiral of 15 millimètres length, and 6 millimètres inner aperture; so that, the slate being shut, a slate-pencil can be stuck through both these spirals, by which means the two slates can then be firmly closed together. When thus closed, the space covered by these two spirals on the front side in the middle of the wooden frame amounts, like the hinges, to 40 millimètres, while between the two spirals is still left a small interval of 3 millimètres. On the outside, the slates are cased with brown lacquered wood.

With one of these slates I betook myself, on the 6th May 1878, in the forenoon, to the residence of my colleague Wach, Professor of Criminal Law in this University, and imparted to him my above-mentioned idea. Professor Wach was entirely of my opinion, that such a slate, if firmly sealed after insertion of a small piece of pencil, and then written upon inside in the presence of Slade, would afford convincing proof, even for persons who had not themselves taken part in such a sitting, of the reality of one of the most remarkable phenomena occurring in Slade's presence. My colleague was also ready immediately to make an experiment himself in the manner proposed. After a small splinter of pencil, of the size commonly used by Slade, was laid upon one of the slates, the slate was shut and then fastened by sticking two strips of paper, 35 millimètres broad, with liquid glue over the shorter frame (184 millimètres long). Over the edges of the strips of paper so glued on Professor Wach also placed two seals, on each side, impressed with his own signet. The strips of paper were intentionally inscribed on the inner side to facilitate discovery in the event of an artificial reunion after tearing. My suggestion to place two seals also on the front side for greater security, my colleague rejected as superfluous, since he was firmly convinced that the securing with four seals completely sufficed already for the discovery of any artifice. With the slate thus fastened I repaired to the residence of my friend Oscar von Hoffmann, and told him my design in this way to induce conviction of the reality of some remarkable spiritualistic facts even in those who were not taking part in the sittings. I gave it, at the same time, as my opinion that it would be much more convenient for strong mediums to convince the world of their innocence in this way than by public or private sittings; and that Mr. Slade could render his material existence much less troublesome and full of care by simply allowing such well-sealed slates to be sent to him, for a fixed price, in order to be returned to the sender when written upon. Of course the whole applicability and demonstrative force of the proceeding depends on the presupposition that it must be possible so to secure such a double slate that it would be *impossible* for the cleverest conjurer or other artist so to open and fasten it again, that this operation could not be detected by the sender of the slate on receiving it back. In principle, indeed, the postal authorities and the public go upon this assumption in the transmission of well-sealed letters containing money. Consequently, for the experimental application of this proceeding with respect to Slade's slate-writing, the condition that the so sealed slates should not be accessible to Slade before the sitting, is *eo ipso* dispensed with. For it is just this precaution which is laid aside as superfluous by reason of the secure mode of fastening, so that Mr. Slade, even if he wished fraudulently to open the slate, and after writing on it a message, to close it up again, would find it impossible to do this without discovery. The aim of the whole proceeding would thus fail, if the condition had been imposed upon me to keep the so sealed slates continually under my care and observation till the sitting. After my conversation with Herr Oscar von

Hoffmann, I therefore placed that slate quietly in the room in my friend's house appointed for Slade's use. Slade himself, so far as I can recollect, was not at home at that time ; and I first saw him again on the evening of that day (6th May 1878 at about 8.45) for the purpose of a sitting. After some words of greeting I took the slate from the closet * near the table, and explained to Mr. Slade, who now apparently saw the slate for the first time, the object I had in view in regard to it. We both, one after the other, satisfied ourselves, by shaking, that the small piece of pencil was between the surfaces of the two slates. I now laid this slate on that side of the card-table (to Slade's left) where were the other slates and different objects, with which *it remained lying from now continuously under my eyes.* Immediately after laying down the slate I sat with Slade at the card-table, on which a brightly burning candle stood. Slade hereupon took up again in his hand the slate referred to, I narrowly and continually watching it, and asked me whether I would not like to affix two seals to both sides of the above-described cylindrical brass spirals, and to impress them with my own signet. Having the latter in my pocket, and a stick of sealing-wax lying on the table among other writing utensils, I at once,

---

* In the frontispiece, showing the sitting-room, will be seen the closet, with other objects on it, among the rest a Mitchell's Polarising Saccharometer.

on the above words of Slade, took the slate with my left hand, drew the signet from my right trouser pocket, laid it on the table, then took the sealing-wax, holding the slate all the time with my left hand, with the wooden edges which had to be sealed turned upwards. Thereupon, holding these edges firmly pressed together with my left hand, I placed on the above-indicated places two large seals, on which I pressed my signet. When the wax had become cold, the two wooden edges of the closed slates were thus so tightly connected that it was *impossible* to push a sheet of paper through those parts which were not stuck with paper and seals. Thereupon I laid the slate so fastened upon the table, and indeed at a place at least a foot and a half removed from Slade's hands, which lay under mine, and were thereby controlled. I now joined in conversation with Slade, and asked him, among other things, whether he had not yet tried, instead of slate-writing, to obtain writing with lead pencil and paper, since this would be an extremely interesting variation of the direct writing produced in his presence. Slade replied that he had not, but was at once ready to make the attempt. We unlinked our hands, and I took from the writing utensils lying ready on the table a half sheet of common letter paper (219 millimètres long, 143 millimètres broad, manufacture mark *Bath*), folded it again about the middle, as if it had to be put into a large letter-cover 144 millimètres broad and 110 millimètres deep, and laid between the two halves of this sheet a cylindrical piece of graphite of 5 millimètres length and 1 millimètre thickness, such as is used for lead-pencil holders. I was about to lay this piece of paper, so folded with the bit of graphite lying in the fold, under the above-described sealed slate, when Slade, under control, proposed that I should tear off two bits from a corner of the folded paper and keep these by me. I at once recognised the importance of this precaution, to establish the identity of the piece of paper in case it was written on, or disappeared and reappeared after some time. Two pieces were therefore, according to Slade's suggestion, torn off at the same time from one corner of the folded half sheet, and these I forthwith put into the gold-compartment of my purse. Then the slate was again laid on the above-described place on the table, and under it was pushed the folded half sheet of letter paper with the stick of graphite lying between the folds, so that the slate completely covered it. We next laid our hands again upon the table, as before, Slade's hands firmly covered by mine, and thus prevented from moving.

We had sat quietly in this position for some time, perhaps five minutes, but nothing worth notice occurred. Slade often shuddered, as by a spasm passing through him, but all remained quiet, so that we became impatient, and Slade resorted to his usual expedient of begging information from his spirits, by help of a slate held half under the table. We unjoined our hands for this purpose. Slade took the uppermost of the slates, which always lay in readiness at his left, bit a splinter from a slate-pencil, laid it on the slate, and held the latter with his left hand half under the table, while he placed his right hand again under both of mine. We forthwith distinctly heard writing, and very soon afterwards the three ticks (*tick-tacks*) which announced that the writing was finished. When the slate was drawn out and eagerly examined by us, the following words were upon it— "Look for your paper." I immediately raised the sealed slate to look for the folded sheet of letter-paper pushed under it, with the bit of graphite inside, about five minutes before : both had *disappeared*. I was startled, indeed, at this unexpected phenomenon, but not particularly astonished, since I had already in earlier sittings witnessed the disappearance and re-appearance of objects so abundantly and under such stringent conditions, that this fact in and for itself offered nothing any longer new for me. I looked often anxiously to the ceiling of the room, in the hope that the paper would fall down, by good chance *written upon*, but it came not, nor did anything else remarkable happen. I therefore desired Slade again to ask his spirits in the usual manner, which he at once did by means of one of the slates lying ready.

The noise of writing was immediately heard, and on the slate being withdrawn, *was upon it*—" The paper is between the slates, and *it is written on it*" (*sic*). Highly pleased at the ingenious combination of physical and intellectual phenomena, I forthwith seized the sealed slate, shook it violently, and in fact distinctly heard the shifting movement of a paper lying between the sides. Notwithstanding the lateness of the hour (it was about half-past ten o'clock), I repaired at once to the residence of my colleague Wach, in order that the double-slate, sealed by him in the morning, might be opened in his presence and by himself. However, I did not find Professor Wach at home; I could only leave word that I would come again the next morning. The slate itself I did not let out of my custody, and took it with me to my residence for the night. Previously, I returned to the residence of my friend O. von Hoffmann, and informed him of my fruitless visit to my colleague Wach. We decided to request the latter the next day to go with us (Von Hoffmann and me) all together to the residence of my colleague *Thiersch*, there to open the sealed slate and take a view of the contents. Herr Counsellor Thiersch was so far interested in this experiment, that he likewise had furnished me with a slate, sealed with the greatest care and circumspection, for application to the purpose named. The continuing and advancing phenomena, and the certainty with which they presented themselves in my daily sittings with Slade (morning and evening), for the most part immediately we had taken our places at the table, had so raised my confidence in the success of *all* my proposed experiments, that I engaged without thinking in a wager with my colleague Thiersch to the amount of 300 marks, which in event of failure I engaged to pay in any form he thought proper. On the other hand, if writing appeared within the slates sealed by him, I desired the sum of 300 marks due to me to be paid as recompense to Mr. Slade. My colleague Thiersch accepted this wager, and being fond of a good cigar, proposed that I should pay him the appointed value of the bet in the shape of a thousand cigars. I requested my colleague to send me a slate well sealed to the residence of Herr O. von Hoffmann that evening, when, with my mother, I took supper in company with my friend's family circle and Slade. When the slate in a great sealed packet was brought and delivered to me towards eight o'clock, just as we were sitting at table, I mentioned half jestingly to Mr. Slade the object with which it was sent, also the bet, exclusively in his interest, with my colleague. A certain displeasure was at once apparent in Slade's features, as if I had done something repugnant to his feelings, and for which I had not been authorised by him. I endeavoured to allay his scruples by the remark that I indeed had concluded the wager on my own account, and it rested entirely with me or with him to apply the 300 marks when won to a benevolent object. Slade replied he would very willingly try whether his spirits were ready to write upon that slate, but he refused beforehand acceptance of any of the money in case of success; he begged me, notwithstanding my last remark, to retire from the wager into which I had entered. I accordingly wrote by return a few lines to my colleague Thiersch, informing him of Slade's decidedly expressed wish, and that under these circumstances the agreement between us had fallen through. I have intentionally communicated in some detail this rejection by Slade, as reflecting a trait of his character, in order, on the one hand, to show his opponents the injustice of their allegations that he is a fraudulent conjurer who wishes to make " money " and " commerce " (*Geschäfte*) of his " art ; " on the other, to enable my readers to form a judgment, from the contents of the following slate-message, upon the moral characteristics of Slade's " four-dimensional intelligent beings." The original English text of this communication, obtained upon a slate on the 6th May 1878, is word for word as follows :—

" DEAR FRIENDS,—A work is before you of a vast interest to all humanity, and is the best to follow the plans laid down by us in order to develop the good that is to come out of your investigation—*never* make any boast, or *never* put up money on this holy subject—it is a law not made by men but by God—we will bring you light as fast as you are able to see—and not be blinded by its rays."

When, next morning, I made my appearance again at the residence of my friend O. von Hoffmann with my sealed slate, in which should be the piece of paper written on with pencil, Slade fell suddenly, at breakfast, into one of his well-known trances; and with closed eyes and altered tone of voice made an address to me in English which, in conclusion, contained statements of what we should find—on opening the sealed double-slate—written with pencil on the paper lying therein. As generally in such cases, Herr O. von Hoffmann wrote down, as far as possible, the words spoken by Slade during his state of trance. They were as follows : *—

" Persevere firmly and courageously untroubled about thy opponents, whose daggers drawn upon thee will turn back upon themselves. The scattered seed will find a good soil : the minds of good men, although lower natures are not able to value it. In what you have witnessed, others later on will discover new beauties which escape you at the time. For science it will be an event of unprecedented significance. We rejoice that the atmospheric conditions have been

* The original English is not given, so that the German translation has to be re-translated into English, not, probably, verbally identical with Slade's language.—*Note by Translator.*

favourable to us, for the conditions must be present, and, in part, prepared. They cannot be explained any more than those, for example, which must immediately precede the falling asleep. Neither in the one case nor in the other can they be compelled. Many enemies of the movement will be its friends, as one of the most important—Carpenter, whose antagonistic disposition has been already, now, through thy labours, somewhat shaken, and who later will be thy fellow-labourer in the same field. *As regards the manifestation of yesterday evening, you will find upon the paper sentences in three different languages; there are some faults in the German and English. At the lower end you will find circles, by which we will denote the different dimensions of space.* To-morrow morning O. von Hoffmann shall again take part in the sitting, and to-morrow evening something strange will happen." *

These words were spoken by Mr. Slade, in a trance, as remarked, somewhere about ten o'clock in the morning of the 7th May 1878, quite unexpectedly to us, during a lunch-breakfast, and three hours later I met my colleagues Wach and Herr O. von Hoffmann at the residence of the Counsellor Thiersch, in order to open the slates fastened with six seals, and which had been up to this time continually in my custody. When this was done, we found, within, the piece of

* On the evening of the 8th May (from 8.20 to 8.35 o'clock) the two endless leather strips were knotted fourfold under my hands, held over them. See *ante*, p. 81.

paper which had been folded by me the evening before, with the stick of graphite, *completely smooth*, without showing any other foldings whatever which could denote a forcible insertion through a narrow cleft. This would moreover have been altogether impossible without injury to the seals, since the extent of the edges of the frame left free between my seals and the strips of paper employed for fastening by Professor Wach—quite apart from their tight adhesion to each other—amounted at the maximum to only 80 millimètres, whereas the narrowest side of the folded sheet of letter paper amounted to 119 millimètres. The often-mentioned two brass spirals on the front side of the slate clasped one over the other in such a manner that every possibility was excluded of shoving in a piece of paper from this side. After opening the slate, I took from my purse the two bits of paper torn off on the evening before and satisfied myself and my friends of their perfect adaptation to the sheet of paper found. All little irregularities of the edges fitted into each other so exactly, that not the slightest doubt could prevail that the torn-off bits of paper formed the completion of the half sheet of letter paper.

I reproduce here the writing obtained, so far as it is possible for me to read it.

Gottes Vatertreue geht
Ueber alle Welt hinaus
Bete dass sie (?) kehrt
Ein in unser armes Haus

Wir mussen alle sterben
Ob arm wir oder reich
Und werden einst erwerben
Das schöne Himmelreich.

Now, is the 4th dimension proven? We are not working with the slate-pencil or on the slate, as our powers are now in other directions.

The strange writing is unknown to me. (Javanese?)

Thus was fully established the correctness of that which Slade had said in the state of trance about the contents of the writing three hours before opening the slate. If I had not had the sealed slate from the end of the sitting continually in my custody, it would be possible, by disregarding the circumstances described by me above with the utmost exactitude, under which the sheet of paper disappeared and was written upon, just on this account to raise suspicion against Slade, as was in fact the case with my colleagues Thiersch and Wach. Already the circumstance that the writing was not, as expected, with the slate-pencil on the inside of the sealed slate had awakened their distrust, and was looked upon by them as a violation of the conditions prescribed by them. I myself, who had personally witnessed all the above-described manifestations, and was accustomed to similar deviations, was exceedingly pleased with the result obtained. It was also in fact far more instructive for me than slate-writing produced between the slates would have been.

For of the reality of *that* fact I had satisfied myself* so often, and under such stringent conditions, partly alone, partly with my honoured friend W. Weber that I myself could have learned absolutely nothing new thereby. On the other hand, through the modification of the experiment, first, my wish was fulfilled of getting writing with lead-pencil upon paper instead of on a slate; secondly, I obtained a splendid proof of the apparent penetration of matter; thirdly, an equally cogent proof of clairvoyance, since Slade, to whom nothing of the contents of the sealed slate could be conveyed by his senses, was nevertheless able to make a correct statement concerning them in his state of trance.

This admirable economy of instruction, which is evidenced in the whole arrangement and progress of the phenomena that I was so fortunate as to observe in Slade's presence, proves for me, more than all other circumstances, the high intelligence and friendly disposition of those invisible beings, under whose guidance these experiments were.

I can here only thankfully express that conviction by again referring to the comparison already made between these *unexpected* occurrences and the providential fatality observed in life.†

---

* Compare the experiment in presence of W. Weber (described *ante*, p. 44), in which a long writing was obtained between two slates bound together crosswise, not touched by Slade's or our hands—these all lying linked together on the table.

† See *ante*, p. 101. The passage is repeated in the text from a former volume. The phrase "providential fatality" is not the author's, but appears to summarise the view expressed in the passage referred to.—Tr.

Chapter Eleventh.

WRITING THROUGH A TABLE—A TEST IN SLATE-WRITING CONCLUSIVELY
DISPROVING SLADE'S AGENCY.

THE most physically astonishing thing in the experiments hitherto related, is, without doubt, the facility with which material bodies apparently pass through each other. Thus, the folded sheet of paper, without betraying the slightest traces of force applied, or of pressure in the transit, had apparently penetrated through a slate covered outside with wood into the interior of the double slate.

I obtained one of the most remarkable confirmations of this apparent suspension of the law of impenetrability of matter in a sitting on the 9th May, 1878, from eleven to a quarter-past eleven in the morning. Immediately after I had sat down with Slade at the card-table, I conversed with him at first on the power of his invisible intelligent beings, by means of which material bodies could be *apparently* penetrated with as much facility as if they were permeable. Slade shared my amazement, assuring me that never until now had such an abundance of this sort of phenomena been observed in his presence. Immediately after this remark he took up with his left hand two slates of equal size from among the slates which lay on the table at his left, and which had been bought and cleaned by myself. He handed me these two slates, and desired me to press the one upon the upper surface, the other against the under surface of the table, with my left hand, so that the thumb of my left hand pressed the upper, my other four fingers the under slate, against the flat of the table, as may be seen from the woodcut, Plate VII. Beneath the upper slate on the table, a splinter of slate-pencil had first been laid, so that it was thus completely covered by the upper slate. Slade then placed both his hands on the middle of the table, about a foot from the two slates, and requested me to cover his hands with my right hand. Scarcely was this done when I distinctly heard writing on one of the slates which were pressed firmly by me against the table. After the conclusion of the writing was signified, as usual, by three ticks quickly in succession, I took the slates apart, and of course expected that the one which had been above the table would be that written on, since on the table still lay the bit of pencil in the same place in which I had laid it a minute before. How great was my astonishment to find the under slate written on, on the side that had been turned to the table. Just as if the bit of pencil had written through the three-quarter inch of oak table, or as if the latter had, for the invisible

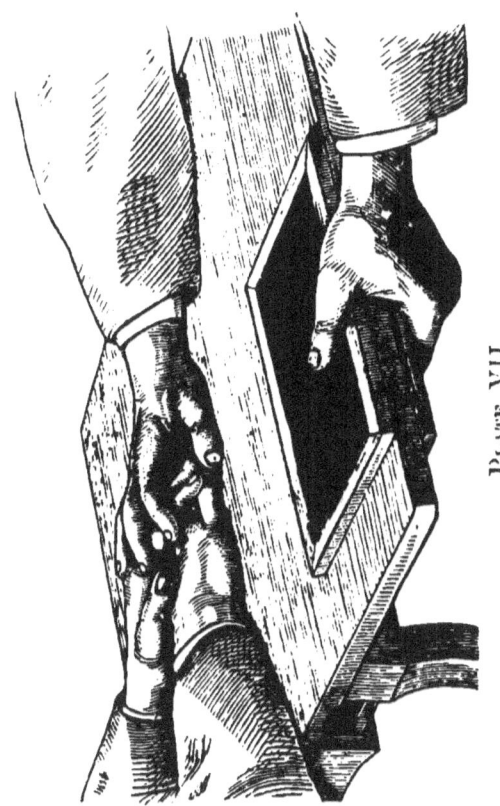

Plate VII.

writer, not been there at all. Upon the slate was the following message in English:—

"We shall not do much for you this morning—we wish to replenish your strength for this evening; you will require to be very passive or we shall not be able to accomplish our work.

"The table does not hinder us the least—we could write in this way more often, but people are not prepared for it."

On the evening of the same day (6th May, 1878) took place the amazing transport of the wooden rings from a sealed string of catgut to the foot of a wooden table.*

In order to meet the above suggestion, so repeatedly raised, that Mr. Slade himself writes on slates by means of a small piece of pencil which he has inserted between the nail and the flesh of one of his fingers, I had purchased at the above-mentioned writing-utensil-warehouse of Mylius, half a dozen slates, of such dimensions that such a manipulation was absolutely impossible. I here presuppose in my readers so much understanding, that they concede to me that any one who will write on a slate in the manner indicated, while holding it at the same time, must be able to touch with his fingers all those parts of the slate which are written upon. Now the slates purchased by me possess a length of 334 millimètres, and a

* *Ante*, p. 105.

breadth of 155 millimètres, with the manufacture mark A. W. Faber, No. 39. Grasp and hold such a slate as one will, even the largest human hand with the fingers completely spread out, cannot by a long way reach all points of the slate-surface. Is therefore such a slate, in the way usually employed by Mr. Slade, written over upon its *whole* upper surface, so is the above-adduced explanation *physically* impossible, and therefore out of the question.

When I repaired with Slade to our sitting-room at the house of my friend O. von Hoffmann at half-past eight on the evening of the 7th May, 1878, I took with me several of such slates, bought by myself, and first carefully cleaned, and laid them down before me on the card-table, at which we at once took our places. Scarcely were we seated, when Slade fell into a trance, which till then had never happened so immediately after our sitting down, folded his hands, and uttered, with altered voice and head upturned, so fine a prayer, that I never shall forget the impression which the noble speech and the fervour with which the prayer was spoken made upon me. The impression was to me so unexpected, and interested me, by the æsthetic in the whole demeanour of Slade with his almost transfigured countenance, so highly, that I did not remember to write down the words. The substance of the prayer was a petition to God further to vouchsafe His blessing on our experiments, and to suffer the work undertaken to end happily for the good of mankind. As usual with Slade, on waking out of such states of trance, there was first a rolling motion of the head, and then he awoke suddenly with a spasm, which shook his whole body, and there was always, before opening of the eyes, a peculiar cracking of the muscles of his neck and jaw. Of what he had spoken in trance, Mr. Slade asserted that he knew absolutely nothing. Those who have been witnesses of the experiments of the magnetiser Hausen will be able most clearly to represent to themselves the demeanour of awaking out of these trance-states, if they recal the expression of the "sensitive" at the summons—"Awake!" of the magnetiser.

After Slade had awoke, his glance fell upon the newly-added oblong slates. His question, for what purpose they were designed, I answered evasively. Hereupon he proposed to try again whether spontaneous writing would be produced upon two slates laid one over the other, not touched by either him or me, as in the experiment which had succeeded so splendidly in the presence of William Weber and me on the 13th December 1877, when between two slates bound together crosswise with strong pack-thread, and which lay quietly on the card-table, neither Slade nor we touching it, a writing was suddenly produced, perceptible to us all.*

* See *ante*, p. 44.

Slade now desired me to take two of the new slates, to lay a splinter of slate-pencil between them, and then to seal these two slates firmly together. I did this, after having again satisfied myself that the slates were perfectly clean. The sealing was in four places on the long sides, and now I laid these slates, with the bit of pencil between them, on the corner of the card-table most remote from our hands. The latter we joined over one another on the table, so that Slade's hands were covered by mine, and were thus prevented from moving. Scarcely had this happened when the untouched slates were raised many times upon one of the edges, which was clearly perceived by us both by the bright light diffused by a candle standing on the middle of the card-table. Then the two slates laid themselves down again on the card-table in a somewhat altered position, and now writing between the slates began to be immediately audible, as if with a slate-pencil guided by a firm hand. After the well-known three ticks had announced the conclusion of the writing, we sundered our hands, which up to this time had been continually and firmly joined, closed the sitting, and betook ourselves with the double-slate, which I had immediately seized, to the next room, where Herr von Hoffmann and his wife awaited us. In presence of these persons the slate shortly before sealed by me was opened. *Both* sides were completely written over in English. (Plate VIII.

PLATE VIII.

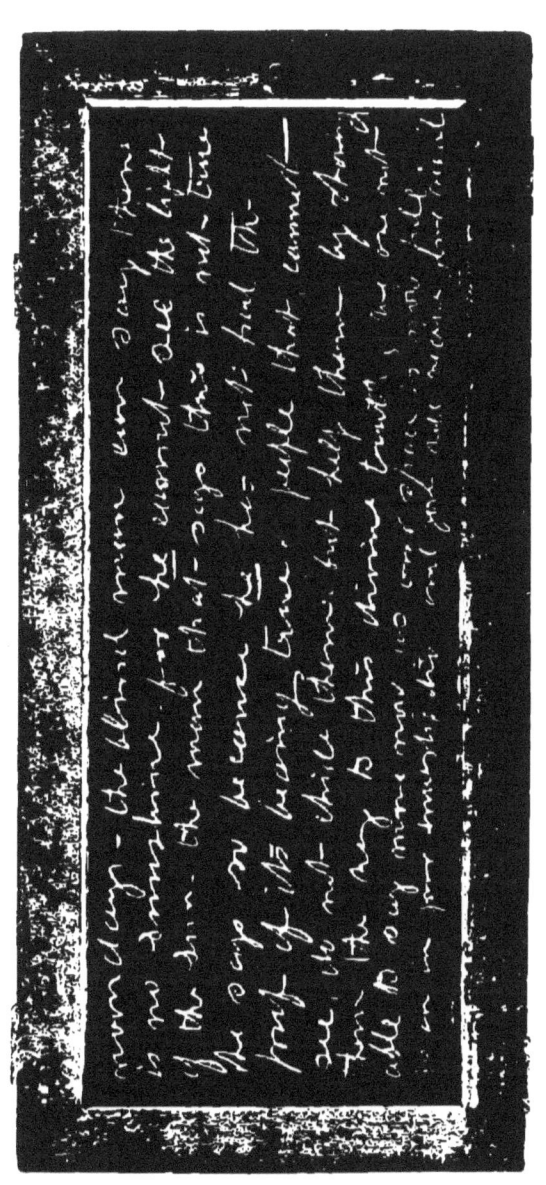

(*Copied from a Photograph.*)

represents a photographic reproduction of the two slates in reduced scale. By clapping together the slates, the two inscribed sides, lying one over the other, show the position in which these surfaces were in fact written over.) Here follows the English original.

"This is a truth—not for select—but for all mankind—without respect of rank or race—no matter how one may be insulted or persecuted by his investigation—it will not take from them the truth, no more than a blind man's words ; by saying there is no sunshine, it does not prevent the sun from shining or bring darkness at noonday; the blind man can say there is no sunshine, for he cannot see the light of the sun. The man that says this is not true, he says so because *he* has not had proof of its being true ; people that cannot see, do not chide them, but help them, by showing them the way to this divine truth ; we are not able to say more now as our space is now full; go on in your investigation and you will receive your reward."

Chapter Twelfth.

A "FAULT" IN THE CABLE—A JET OF WATER—SMOKE—"FIRE EVERYWHERE"—ABNORMAL SHADOWS—EXPLANATION UPON THE HYPOTHESIS OF THE FOURTH DIMENSION—A SÉANCE IN DIM LIGHT—MOVEMENT OF OBJECTS—A LUMINOUS BODY.

I PASS now to the account of further facts observed by me, which prove the intimate connection of another material world with our own, and may be considered in general as a confirmation of the numerous observations of Mr. Crookes and other physicists. Generally, hitherto, my accounts have had reference to the sudden disappearance and return of *solid* bodies; the following facts will show the advent (*Eintritt*) of bodies in the fluid and gaseous condition, without our being able, from the standpoint of our *ordinary* and *limited* conception of space, to give an answer to the question, "whence?"

On the 7th May, 1878, at fifteen minutes past eleven in the morning, I had taken my place with Slade at our card-table. In order that we might first learn something of what we were to expect, I took one of the slates kept in readiness, cleaned it, laid a small bit of slate-pencil upon it, and handed it to Slade to hold, as usual, half under the edge of the table, that it might be written on by his invisible beings. Slade proposed, as a variety in this proceeding, the following modification. He desired me to press the slate from below against the table with my left hand, as is shown in the above wood-cut,* while he grasped the slate at the other corner with his right hand, and pressed it in the same manner against the table. His left hand Slade laid extended on the middle of the table, and I covered it with my right hand. Scarcely was this done when writing began on the slate; this gave the opportunity of confirming a phenomenon observed also at other times by myself, and frequently by others, that the distinctly audible sound of writing immediately ceased as soon as by raising my right hand I removed it somewhat from Slade's left. As soon as the connection was re-established, the writing immediately recommenced.† Three ticks on the surface of the slate having declared the writing ended, the following was found on the slate on its upper surface, which had been pressed against the table :—

"To-morrow morning we would be pleased to have Baron H. sit with you—and shall begin a new power, and give you more proof of what can be done; please

* Ante, p. 193.
† The translator observed this on several occasions when sitting with Slade in London, in 1876. The same fact is also recorded by the late Mr. Serjeant Cox in the "Spiritualist," August 1876.--TR.

ask us no question, or make any more requests ; we will do all in our power for you—we wish to say more to-morrow morning by controlling the medium."

Slade and I then rose to look in a closet near for a somewhat larger piece of slate-pencil, but before this could be done, almost in the moment when we rose, we were sprinkled from above by a sort of drizzle. We were both wet on the head, clothes and hands, and the traces of this shower—of perhaps one-fourth of a second duration—were afterwards clearly perceptible on the floor of the room.

Remains of the liquid being especially on the upper side of my right hand, I touched it with the tip of my tongue ; so far as taste could inform, the moisture was pure water. I should mention here, that in the room in which we were there was no vessel with water, although there was in that immediately adjoining. After the above-related facts concerning the transport of solid bodies from three-dimensionally enclosed spaces, such a conveyance of water from one room to another would appear to be a phenomenon of the same kind.

Surprised at this unexpected phenomenon, and yet busied in drying our clothes, we took our places again at the table, and were about to join hands, when suddenly the same thing was repeated almost more strongly. This time the ceiling and walls of the room were also moistened, and there seemed, judging from the direction and form of the traces of water, to have proceeded several different jets of water at the same time from a point in the middle of the room, perhaps four feet high above our heads; as if a jet of water were to be discharged perpendicularly upon a plane, where it would then spread itself out radially in all directions in this two-dimensional region of space, from the point at which it reached the ground. If one applies this analogy to a jet of water discharged from the fourth dimension into the three-dimensional region of space, the water would then appear at a particular spot of this space, and under suitable conditions must extend itself thence radially to all three dimensions.

I may further remark that I met with the same phenomenon in just as unexpected a manner, at a sitting with Slade, at which Herr Gillis of St. Petersburg was present.* Since that sitting took place in the sitting-room of the restaurant-keeper of the Thuringian railway station, which Slade had set foot in shortly before for the first time, the possibility of a conjuring apparatus is excluded; and independently of this, the same phenomenon in the presence of Slade has since been confirmed by numerous other observers.

On the next morning (8th May 1878, eleven o'clock), Herr O. von Hoffmann took part in the

* Referred to, but not described, in an earlier part of the volume.—TR.

sitting; he sat at my right, Slade in his usual place at my left. After some short writings on the slate had been obtained in the usual manner, and Slade had joined his hands with ours again on the middle of the table, there rose suddenly, in three different places above the edge of the table from beneath, a smoke, which, judging from the smell, contained some acid of sulphur and saltpetre. We immediately looked under the table, but saw nothing further than the still present remains of this smoke, as after the lighting of a lucifer match. Scarcely had we again joined our hands, to await the further development of the phenomena, when the same thing was repeated yet more strongly.

Almost at the same time Slade proposed to me to place a candle under the table, to see if the invisible beings were able to light it. Thereupon Herr von Hoffmann took two candlesticks, provided with new unused candles, from his writing-table, and placed them both on the floor under the table,* at which we immediately resumed our seats, and joined our hands in the manner already mentioned. After we had waited for some minutes, smoke rose up again from under the table, almost from all sides; and at the same time one of the candlesticks with the candle burning hovered up above the edge of the table

---

* Not under the middle of the table, but under the edge at his right, the place furthest removed from Slade's feet.

opposite to me; after a few seconds it sunk down again; and when we looked under the table one of the candles was lighted, and under the middle of the table. To refute the suggestion of a transient hallucination or "unconscious cerebration," a half sheet of writing-paper was taken, held close over the burning candle, and in this way a hole was burned through the paper. I then took a stick of sealing-wax, held it in the *same* light, and let a part of the melted wax drop on the paper, and then impressed the seal with my signet. The half sheet of paper with the seal under the in-burnt hole is still safe in my possession. After our agreeable astonishment at this unexpectedly successful experiment had somewhat subsided, we sat again at the card-table, and placed the burning light in the middle of it. Scarcely had this been done when Slade fell into a trance, and with closed eyes uttered an address, of which Herr von Hoffmann took down the following words, while Slade was slowly speaking them :—

"All seems strange that is not understood; fire is everywhere. Think of the flint from which you draw it; it is in all the elements around you. Let this light be a beacon light in the path of investigation, let it be symbolical of the light that must break through the darkness of the world. The light of the brain will light thy pathway! This evening we will enter into a new phase; to-morrow morning we will replenish the forces, and in the evening show you another phase, if the atmosphere be favourable."

In fact, our invisible friends kept their promise of the morning in a manner astonishing to us all.

We were sitting at half-past seven in the evening, at the tea-table in the dining-room. On the table burned a large lamp; Slade sat opposite me, his back turned to the window, the curtains of which were let down. At my left, on one side of the table, sat Frau von Hoffmann; opposite to her, on the other side of the large tea-table, Herr von Hoffmann; I myself had my back turned to the great folding-doors, provided with a brown curtain, by which one entered the room from the corridor. Since in general we had never observed remarkable manifestations with Slade during meals—I leave quite out of sight particular risings of the table and movement of detached chairs—we naturally were not expecting anything surprising on this evening. Suddenly, however, Frau von Hoffmann cried out, and said that she saw on the wall and on the door to which my back was turned the reflection of a bright light which appeared to issue from a place under the table at which we sat. Slade, who from his place was facing the side of the room referred to, confirmed this assertion. We looked first under the table, examined everything narrowly, but found nothing which could explain the origin of such a light. In the expectation that this phenomenon would perhaps be repeated, we frequently looked at the side in question, and for easier observation I had placed my chair somewhat obliquely. Suddenly this phenomenon occurred again, and then, immediately afterwards again. The colour of the light was bluish-white, as if proceeding from a suddenly kindled electrical light, and, what was for me the most remarkable, the shadows of the feet of the table were sharply projected, nevertheless, so far as I could ascertain in the short time, *perceptibly of the same size as the objects casting the shadows*. Although I might consider this phenomenon, owing to the want of sufficient test-conditions (*Controle*), as a *not scientifically* established fact, raised above all doubt, yet I hold it, nevertheless, for my scientific duty to make mention of it, in order that other observers may be attentive to its extraordinary importance.

If, for instance, the origin of this ray was a luminous point in the space beneath the table, the shadows of the feet of the table must, according to the laws of shadow-casting, have been considerably larger on the wall than the feet themselves, as any-one can easily prove by placing a lighted candle under a table having several feet. The size and form of the shadow-projection of an object approximates, as one knows, more to the size of the shadow-casting object, the further the source of light is removed from the latter, or in other words, the nearer the rays are to the parallel. The *sharpness* in the outline of the shadow affords, moreover, an inference as to the apparent size of the light-source; if, for example, the apparent diameter of the sun's disc were twenty times greater than is in fact the case, the shadows cast by opaque bodies in sunlight would be effaced at the edges to a far greater extent than actually happens. Apart from the phenomena of refraction, a body would cast an absolutely sharp shadow of absolutely similar size with the shadow-casting object, if the rays proceeded from an *infinitely* remote point. Since, now, in the above-mentioned case, surprisingly sharp shadows of the feet of the table *of perceptibly similar size* to the feet themselves were observed, it follows from this that the rays which produced that projection of shadow, must have issued from a light source, *first*, possessing a very small apparent size, and, *secondly*, being at a great distance. No place *underneath* the table could have satisfied the second condition, and since the remaining space of the room was observed, and even the distance of the nearest wall at Slade's back, would not have sufficed to comply with the above condition, the said phenomenon would thus point to another place as the point of issue, which cannot lie at all in our three-dimensional space. This contradiction is solved as soon as one presupposes the reality of a four-dimensional region of space, and admits that it is possible for those invisible intelligent beings, who have showed us so much of their powers, also to divert rays of light, which are diffused in the direction of the fourth dimension, so that they fall in our three-dimensional region of space. We are, indeed, likewise able, by reflection and refraction of light, to divert rays in such a manner as to transfer their point of issue to another than the true place. Upon this diversion of rays of light depends most of the physical-optical illusions. Since similar phenomena of lights are very frequently observed at spiritualistic sittings, and among others, Mr. Crookes has also given detailed testimony to them,* I may be

* "*Notes of an Inquiry into the Phenomena called Spiritual,*" by William Crookes, F.R.S. London, 1874. Mr. Crookes enumerates and describes thirteen classes of phenomena observed and verified by himself in his own house, and with only private friends present, besides the medium. Of Class viii. "Luminous Appearances," he says : "These, being rather faint, generally require the room to be darkened. I need scarcely remind my readers again that, under these circumstances, I have taken proper precautions to avoid being imposed upon by phosphorised oil, or other means. Moreover, many of these lights are such as I have tried to imitate artificially, but cannot.

"Under the strictest test conditions, I have seen a solid self-luminous body, the size and nearly the shape of a turkey's egg, float noiselessly about the room, at one time higher than any one present could reach stand'ng on tip-toe, and then gently descend to the floor. It was visible for more than ten minutes, and before it faded away it struck the table three times with a sound like that of a hard, solid body. During this time the medium was lying back, apparently insensible, in an easy chair.

"I have seen luminous points of light darting about, and settling on the heads of different persons ; I have had questions answered by the flashing of a bright light, a desired number of times in front of my face. I have seen sparks of light rising from the table to the ceiling, and again falling upon the table, striking it with an audible sound. I have had an alpha-

permitted to call the attention of other observers to the circumstance mentioned. For approximate determination of the point of divergence of the rays of such luminous phenomena, the following proceeding may be recommended as the simplest. Phenomena of light are observed by aid of an opera-glass, by the adjustment of which the object may be removed as far as possible. Objects at so short a distance as those in a room, require, to appear in sharp outline, a special adjustment of the glass, and this adjustment— the determinate distance of the eyepiece from the objective—enables us, according to simple optical laws, to determine the distance of the object, that is, of those luminous points, from which the rays extend themselves in space. If, now, it should really appear, with respect to these spiritualistic luminous phenomena, that the distance of the point of divergence of the rays does not agree with the distance of the luminous object, the difference of these two distances would determine the length of a tract (*Strecke*) falling in the fourth dimension, and hereby would be made the first step towards quantitative determinations in

betic communication given by luminous flashes occurring before me in the air, whilst my hand was moving about amongst them. I have seen a luminous cloud floating upwards to a picture. Under the strictest test conditions, I have more than once had a solid, self-luminous, crystalline body placed in my hand by a hand which did not belong to any person in the room. *In the light* I have seen a luminous cloud hover over a heliotrope on a side table, break a sprig off, and carry the sprig to a lady ; and on some occasions I have seen a similar luminous cloud visibly condense to the form of a hand and carry small objects about."

the four-dimensional field of space. Such an observation would, in the history of transcendental physics, be comparable to the first determination of parallaxes in the history of astronomy, whereby we obtained the first approximate conception of the distance of our moon, the nearest to us of the heavenly bodies.

I may mention, that the above-described luminous phenomena were repeated on two other evenings (9th May, and 19th May) under similar circumstances, and in presence of others who were sitting at tea at the same table. On these evenings, however, for the sake of a better control over Slade, and for more convenient observation of the shadow-projection on the opposite side, I had taken my place close beside Slade, so that he sat at my left. The only difference of the phenomenon from that observed on the first evening consisted in the colour of the light being yellowish-red, instead of a bluish-white. It will therefore be useful in future at similar sittings to have with one a pocket spectroscope, to examine the nature of the light, as opportunity offers.

Finally, I mention here a sitting with Slade which took place at five o'clock in the afternoon of the 15th December, 1877, in the usual sitting-room of the house of my friend O. von Hoffmann, whose wife was present. It was the only one in which the room was partially darkened, to try whether in Slade's presence, as in that of the young lady of fifteen (Miss

Cook), a human form, or at least a "phantom form," as Mr. Crookes describes it in his book, under the heading "Phantom Forms and Faces," would be evolved. In order to improvise a cabinet, a string was drawn obliquely across the part of the room opposite my usual place, at about two metres* above the floor, and of a breadth corresponding to that of the edge of the table, a dark green curtain being fixed to it. Slade sat at his usual place, at his right Frau von Hoffmann, I next, and Herr von Hoffmann at my right. We had already laid our hands, linked together, on the table, when I remarked it was a pity we had forgotten to place a small hand-bell on the table. At the same moment it began ringing in the corner of the room at my right front, at least two metres from the middle of the table; and the room being faintly illuminated by gaslight from the street, we saw a small hand-bell slowly hover down from the stand on which it stood, lay itself down on the carpet of the floor, and move itself forward by jerks, till it got under our table. Here immediately it began ringing in the most lively manner, and while we kept our hands joined together as above described on the table, a hand suddenly appeared through an opening in the middle of the curtain with the bell, which it placed on the middle of the table in front of us. I hereupon expressed the wish to be allowed

* About 6½ feet.—Tr.

to hold that hand once firmly in my own. I had scarcely said this, when the hand appeared again out of the opening, and now, while with the palm of my left hand I covered and held fast both Slade's hands, with my right I seized the hand protruded from the opening, and thus shook hands with a friend from the other world. It had quite a living warmth, and returned my pressure heartily. After letting go the hand, I reached it a slate and challenged it to a small proof of strength; I would pull to one side and it should pull to the other, and we would see which of us kept the slate. This was done, and in the frequent give-and-take, I had quite the feeling of an elastic tug, as though a man had hold of the slate at the other side. By a strong wrench I got possession of it. I again remark that during all these proceedings Mr. Slade sat quietly before us, both his hands being covered and detained by my left hand and by the hands of the two others.

I may here point out that such a pull on one side by a human hand or other solid body, as a slate, would be a violation of the principle of the equality of action and reaction, if no material object undergoing the equal, but resisted, pull were to be found in three-dimensional space. But no such object being to be found in the space ordinarily perceivable by us (*in unserem gewöhnlichen Anschauungsraum*), it must occupy a position in absolute space, falling in the

P

next higher region of space. Only in this manner can the apparent contradiction, here introduced, of a fundamental law of the interaction of bodies, be satisfactorily solved for our understanding.

While I was still busied with the above observations and experiments, there suddenly emerged from above the upper border of the curtain, a half circular mass gleaming in phosphorescent light, of the size of a human head. It moved slowly to and fro at the same height from one side of the curtain to the other frequently; and gave us all the impression of appertaining to a luminous form close behind the curtain. Approaching that side of the curtain at which Slade sat, this luminous form became visible in its whole extent. Slade drew back, evidently alarmed, whereat we laughed, and the form immediately hovered back behind the curtain, and with the same speed moved to the other side, here also emerging up to the middle. We could not distinguish features or limbs. In brightness and colour the phosphorescent light resembled that observed in the so-called "after shining" Geissler's tubes. I much regretted that I had not at hand my pocket-spectroscope, in order more closely to examine the nature of the emitted light.

Chapter Thirteenth.

PHENOMENA DESCRIBED BY OTHERS.

THE foregoing comprises in essentials all the phenomena which I have myself observed in Slade's presence during a series of more than thirty sittings and other meetings. The precautionary measures which I had taken on these occasions were such, that for my understanding every possibility of deception or subjective illusion was excluded. I do not, however, assert that these measures will be regarded as sufficient by the understanding of other men. I am therefore quite ready and willing to receive instruction and enlightenment as to better precautions than those adopted by me; provided that my advisers have given other proofs of intellectual competence superior to my own, to induce me to defer to them and to recognise them as judges of facts of observation, which they have not seen, but have learned for the first time from my description.

Before Mr. Slade left Germany, he visited Annathal in Bohemia, by special invitation from Herr J. E. Schmid, the owner of a factory there. In the family of this gentleman he found the most friendly reception, and remained a week. Herr Schmid has already published a short account in a letter to *Psychische Studien* (July 1878). For the following detailed description I am indebted to Herr Heinrich Gossmann, Herr Schmid's bookkeeper, who witnessed all the phenomena during Slade's residence with Herr Schmid, and gave me a verbal account of them when on a visit to Leipsic. In accordance with my request, and by permission of Herr Schmid, he afterwards furnished me the following written account.*

"Mr. Slade arrived here on the 14th May, last year (1878), but was too tired by his journey to give us a sitting on that day. Notwithstanding which, to the surprise of us all, on his entering the room, we heard thundering blows on the sofa, for which Mr. Slade could certainly have made no preparations, as he had never been in the room before. To the question whether this was a manifestation, Mr. Slade replied in the affirmative, remarking that the spirits could not wait till the next day to announce themselves, and that he had often found this to be the case where harmony prevailed. We took our seats at the table, without intending a regular sitting, and had scarcely done so when all at once a seat at some distance, near the piano, put itself in motion, and

---

* The introductory and concluding parts of this letter are here omitted, as not material. TR.

came up to the table of its own accord. Continually as our astonishment increased, we did not neglect to watch Mr. Slade closely and attentively. I was sitting next him, and after some time was swiftly and unexpectedly swung round in a half circle, with the chair on which I sat, so that I nearly fell off it. Others at the table were now touched, sometimes softly, sometimes powerfully, and to me this happened often. . .

"One manifestation now followed another, chairs moved up to the table, touches on our knees were constantly felt, a knife and fork were put across each other on a cloth at the lower end of the table, as if they were cutting meat, then from another side of the table a fork flew off on to the floor in a slight curve.

"On the next and two following days *séances* were held in another room at a table appropriated to them. Many persons, sceptics and the like, to whom Spiritualism was as yet unknown, took part in them. A chain was formed, and we gave Mr. Slade a slate which he had never had in his hands before. He laid on it a small bit of pencil, and asked the spirit of his deceased wife to tell them, by direct writing, if it was possible for any of the departed relatives of the family to communicate in the same way; to which an affirmative answer was returned. Mr. Slade now put the pencil on the table, showed us that the slate was quite clean and without writing, and then laid it on the table over the pencil. Writing under the slate was at once heard; we could distinctly follow the scribbling and taking off of the pencil. This sitting, as all the rest, was in bright daylight; the slate lay there free, before all our eyes, when we formed the chain, and Slade laid one hand on the slate. The conclusion of the spirit-writing was denoted by three sharp raps; and the slate being lifted up, we found the whole under side of it written over, first by an address from Slade's wife in English, and next by a message in German from a spirit-relative. A communication from the deceased father of the lady of the house was especially striking, as his characteristics and habitual expressions when on earth were quite distinctly recognisable in it. Besides the great resemblance of the writing on the slate to that of the deceased, his identity was apparent from a certain manner of speech, and such phrases as 'We must all die,' which came upon the slate. And in many of these communications the like resemblances were observable. Among others, the brother of the lady of the house communicated, *and in verse*, a custom he had when on earth, especially in writing to his sister, whom he generally addressed in rhymes. She recognised her brother very clearly in this, and on comparing the writing with that of his letters, just the same strokes were found in them.

This communication was obtained in the following manner:—

"A young lady (a relative of the family) who sat at the lower end of the table, opposite Mr. Slade, took in her left hand, by his direction, two slates connected by hinges; a small pencil was laid between them, and she joined her right hand to the chain of hands on the table. Mr. Slade sat quite away from the slates, and his hands were likewise joined in the chain; and under these conditions, to our great astonishment, writing began between the slates. The young lady, according to Mr. Slade, was mediumistic, therefore it was that she could obtain writing while holding the slate herself alone, which was not the case with the others; she also perceived the pressure upon the under side of the slate while it was being written upon. . . .

"Such direct writings covered at least twelve slates, which were bought here, and came to Mr. Slade's hands for the first time, before all eyes, without his having any possible opportunity for "preparing" them, or for writing upon them without continual observation. Mr. Slade often held the slate quite sloping, at an oblique angle, and yet the pencil upon it did not slip to the edge, but wrote quietly on. The supposition one so often hears that the slates are "prepared" by Mr. Slade will not stand examination, because he washes out the answers, given to his questions by the spirits, on the slate, which (the same one) is again written upon; this also, as always, happening under observation. When once during a *séance*, at which writing was going on under a slate, one of the circle raised his hand quietly and without being observed, from that of his neighbour, the writing suddenly ceased, the connection being thus disturbed. Mr. Slade looked up, and seeing what had happened, requested the gentleman referred to, to try the experiment frequently, and each time the writing ceased, and began again as soon as the chain was re-closed. There were many other manifestations. For instance, a bell under the table came out of its own accord, ringing, rose high up in the air, and let itself gently down, still ringing, on the table. A slate placed under the table was shivered into small pieces, as by lightning, and the fragments flew in curve over our heads and so on to the floor. During a *séance*, another heavy table which stood at some distance from the one at which we sat, came with a rush of extraordinary speed and force to the side of a gentleman among us, whom we thought must have been hurt; but it only touched him quite gently. The spirits gave to a hydropathic doctor, who was present, a token of esteem for his practice by wetting him with a jet of water, which came from a corner of the ceiling opposite him. Just afterwards my knee was tightly grasped by a wet hand, so that I felt the wet fingers sharply, and on examination I found the moisture on my trouser. (Mr. Slade, during this, had his hands linked in the chain formed by those of all present.)

"Another interesting fact is, that when my Principal (Herr Schmid), Mr. Slade, and I, were holding our hands lightly on the table, the latter went up, hovering in the air, and turned itself over above our heads, so that its legs were turned upwards.

"What an enormous force Mr. Slade must have applied to evoke these manifestations deceptively, is shown by the following case. When I was sitting, a little distance from him, he likewise sitting, he stretched out his arm, and laid his hand on the back of my chair. All at once I was raised, with the chair, swaying in the air about a foot high, as if drawn up by a pulley, without any exertion whatever by Slade, who simply raised his hand, the chair following it as if it were a magnet. This experiment was often repeated with others.

"Mr. Slade held an accordion under the table, grasping it by the strap at the side ; his other hand lay on the table. Immediately we heard the falling-boards move, and a fine melody was played.

"The experiment with two compasses was also tried ; these were placed close together, and when Mr. Slade held his hand over them, the magnetic needle in *one* of the compasses began quickly swinging round in complete rotations, while the needle in the other compass remained at rest, and so also conversely. According to the laws of physics known hitherto, if Mr. Slade had been secretly applying a magnet, as is so frequently alleged by opponents, *both* needles must have been set in motion, as they were quite close together, yet this was *not* the case.

"One of the most wonderful manifestations was the following :—Mr. Slade stood in the middle of the room, I on his right, on my right my Principal, and behind us, at the window, stood a young lady. While in this position we were conversing, and my Principal was about to go into the next room to fetch something, a heavy stone, as if originating in the air, fell before all our eyes with a very heavy blow upon the floor, so that a regular hole was made in the latter; the stone fell quite close to my Principal's feet. Immediately afterwards there fell a second stone, the fall of which, as of the first, we saw very distinctly. This did not happen close to Slade, for I and my Principal were both between him and the place.

"Occasionally at a sitting we saw a materialised hand; it would tear the slate forcibly out of Slade's hand under the table; it appeared suddenly at the side of the table, and quickly vanished again; it was a strong hand, quite like one of flesh and blood.

"A slate was regularly wrenched out of my Principal's hand; it then made the round of the table, *hovering free in the air before all eyes*. . . . Slade came here alone without any companion."

Professor Zöllner next refers to the manifestations obtained through Slade at Berlin, of which he had received information from visitors and correspondents. Among the slates which were brought or forwarded to him, was one written upon in six different languages, and which Professor Zöllner ascertained, upon examination, to be free from the "preparation" by artificial means, so often suggested as the probable explanation of the long sentences coming upon apparently clean slates during Slade's *séances*. In this case, moreover, as will be seen, the slate was brought by the investigators, and was never in Slade's custody at all; nor was there the smallest opportunity afforded for effecting an exchange. The correspondent from whom the author received the account was a "Herr Director Liebing," of Berlin, who obtained the details from the owner of the slate, in whose presence it was written upon, with full authority to transmit them to Professor Zöllner for publication, with the slate. Although it would have been preferable to have had the account direct from this gentleman, it appears from the correspondence in the text (which it is not thought necessary to reproduce literally and at length in this translation), that the statement was submitted to him for correction, was in fact corrected by him, and is thus, as here given, in effect his own. He was a Herr Kleeberg, residing at No. 5 Schmied Street, Berlin, and "of a very respectable firm" in that city. He and a friend of his, a "thorough sceptic," took two slates to Slade. One slate was covered by the other, and beyond putting a piece of slate-pencil between them, *Slade never touched them at all.* Herr Kleeberg and his friend then held the two slates, so joined together by their hands, *above the table*, suspended over it, *in full daylight*, and writing at once began. When it was over, and the slates were separated, the lower one was found covered with writing, as shown in Plate IX. One long passage was in English, five short sentences in French, German, Dutch, Greek, and Chinese (the latter according to the judgment of a student of Oriental languages), respectively. They were as follows:—

1. Look about over the great mass of human intelligence and see for what these endowments are given to man. Is it not to unfold (in) the great truths God has embodied in him? Is it not mind that frames your mighty fabrics? the soul that is endowed with powers. Shall he not go on unfolding these powers as God has sent His angels to do? Must man pass his judgments on God's laws that he does not understand? We say no.

2. Es ist mir schmeichelhaft Sie bedienen zu können. (I am proud to be able to serve you.)

3. Que la grâce soit avec vous tous qui êtes en Jesus Christ. Amen. (The grace of God be with you all who are in Jesus Christ. Amen.)

4. Οἱ πονηροι εἰς το κερδος μόνον ἀποβλέπουσιν. (Bad men look only to their own advantage.)

5. Die het zaadije wasdom geeft, En verzadigt al wat leef't. (Who to the seed-corn increase gives, nourishes all that therein lives.)

The last sentence, supposed to be Chinese, was not understood.

APPENDICES.

APPENDIX A.

THE VALUE OF TESTIMONY IN MATTERS EXTRAORDINARY.*

BY CHARLES CARLETON MASSEY.

THE proposition that evidence, to command assent, should be proportioned to the probability or improbability of the fact to be proved, is constantly appealed to as the rational foundation of sceptical or negative judgment. I ask you this evening to come to close quarters with it, to consider what it means, and whether it is legitimately applied. There are perhaps no two words in the language more liable to abuse, or more frequently abused, than probability, and the word expressing that upon which probability is said to be founded, namely, experience : for there is here no question of those definitely-ascertained probabilities which result from the computation of known chances, and which are, therefore, not matters of experience at all. It is by reference to these, however, that we shall have the principle in question most clearly before us. Suppose, for example, evidence of such a character and amount that the chance against its being forthcoming for what is not true is as 5 to 1, and that it is given for an event against which the chance is as 10 to 1, the resulting probability is 2 to 1 against the

* A Paper read before the Psychological Society of Great Britain, on Thursday, June 6th, 1878.

evidence. Now it is said that the inductions from experience afford us a similar, though not equally definite, measure of proportion between the probability of facts and the value of evidence.

And as to a large class of alleged facts, we are met at the outset of our inquiry by the previous question, whether testimony in relation to them has any value whatever? The probability in favour of testimony, even at its best, it is said, can never equal that which results from the uniform negative experience of mankind. Our faith in testimony is based on the same principle of experience, and therefore testimony can never prove a fact which is contrary to a wider induction. This is the extreme application of the principle, as we find it in Hume's celebrated argument against miracles. It is not quite the same, though practically it has the same effect, as that absolute *à priori* denial of the possibility of the facts attested to which few. scientific minds will explicitly commit themselves. It does not say that our inductions as to what is possible, or *in rerum naturâ*, are certain, but that they have a greater force than any testimony which can be adduced against them, which therefore is not entitled even to consideration.

Now, in the first place, I would invite you to consider when it can and when it cannot be said with accuracy that an alleged fact contradicts experience. In one sense, of course, it cannot be accurately said at all. Your experience that contact with fire has always burned you remains unchallenged and uncontradicted by any assertion of mine that on one occasion or on half-a-dozen occasions it has not burned me. But experience is a term used loosely to denote our inductions from experience; and this is the first thing I ask you to mark. What, again, is a fact in relation to experience? If you and I have seen the same object, and you describe it as of one apparent dimension, and I describe it as of another and vastly different apparent dimension, does my experience contradict yours? Not necessarily; for we may have both described the apparent object abstracted from the conditions of distance under which we severally saw it. This tendency to abstract from the context of experience, in other words, to ignore conditions, is just what distinguishes the popular from the scientific conception of a fact. And until we know *all* the conditions under which anything is said to have occurred, we cannot properly speak of it as opposed to our own experience. The next remark I have to make is that, *à priori*, we do not know which of the circumstances attending even the most familiar facts of experience are conditions, and which are entirely irrelevant. Transport yourself to an imagined infancy of experience, and you could not predict from the fact that fire had burned you in one place or time that it would burn you in another, or that it would burn me. Difference of place, time, or person might, for all you could know beforehand, provide entirely new conditions. Now if it was asserted, as in fact it is asserted with regard to a large class of alleged phenomena, that personality, that specialities of human organism do introduce new conditions, resulting in these unusual phenomena under certain other conditions not scientifically known, this would not be and is not to contradict the common experience which, *ex hypothesi*, knows nothing of these exceptional personalities. Bearing in mind, then, that no experience or amount of experience has the least relevance to an alleged fact except under the exact conditions, inclusive and exclusive, of its occurrence, and that we cannot say beforehand what are conditions and what are not, the experience argument, in relation to the phenomena in question, resolves itself into this: that inasmuch as the alleged personalities which, as the one constant element must be regarded as the condition, are exceptional and abnormal, therefore their existence is so improbable that testimony cannot prove it. What is this but to say that the abnormal can never be proved by testimony? Nay, more, that testimony can never make such a provisional and *primâ facie* case, as to justify a reasonable man in seeking for the higher evidence of his own experience; in other words, in investigating for himself? For such a *primâ facie* case is a probable case, and here it is said that the balance of probability is largely against the fact. I am endeavouring to get at the precise point in issue; and I say that the man who exclaims, "Objects moving without physical contact! writing read without eyes! matter passing through matter! writing without hands! these things are opposed to all human experience!"—is talking wildly and loosely. What, if he would condescend to be exact and logical, he really means is that *it is opposed to a negative induction from the absence of experience that individuals should exist who can provide new conditions of physical operation.* But the question is, Is this induction to be regarded as final? And as we are dealing with the experience school solely with its own weapons, let us see what experience says to that. And I should have thought that if there was one induction from experience historically and scientifically valid it was that other inductions from experience—and especially negative inductions—are *not* final. Our widest inductions are precisely those which we make in the infancy of experience and science. Science advances by the discovery of new conditions which limit general rules. What was rejected as abnormal yesterday is found to have a law of its own to-day. In a word, if the widest and highest experience of mankind can afford us a canon of probability it is this—that testimony, otherwise sufficient, to the exceptional, the abnormal, the strange, and the new, is probably true, and not probably false. Set side by side the cases in which new facts of nature have been asserted and proved to be true with the cases in which they have been well asserted and yet disproved, or not proved, and who that is acquainted even superficially with the history of science and discovery would hesitate to say which list affords us the best foundation for an induction?

I submit, then, as the results of the foregoing considerations—1. That testimony to the extraordinary, of which the phenomena referred to may be taken as a type, is falsely opposed to experience. 2. That what it is opposed to is simply a negative induction from the absence of experience. 3. That a more general experience teaches us that such negative inductions cease to be probably true, so soon as they are opposed to testimony of a character sufficient to establish any other fact.

It is a great satisfaction to me to be able to state that since the above was written, I have found the distinction between positive and negative experience, and the character of the inductions from each, very ably and elaborately explained in a long note by Mr. Starkie, in his *Treatise on the Law of Evidence*. I do not quote this note *in extenso*, because I hope the distinction is already obvious to all. Mr. Starkie's observations refer expressly to Hume's principle of incredulity; and he shows, as Mr. A. R. Wallace has also shown, that pushed to its logical consequences that principle would be absolutely fatal to all scientific progress. One could almost imagine the following passage to have been written in prophetic protest against the appeals to Hume by the sceptics who treat with contumely and derision every testimony to the occult phenomena of the present day. "Experience, then, so far from pointing out any unalterable laws of nature, to the exclusion of events or phenomena which have never before been experienced, and which cannot be accounted for by the laws already observed, shows the very contrary, and proves that such new events or phenomena may become the foundation of more enlarged, more general, and therefore more perfect laws." And in the text Mr. Starkie says—" As experience shows that events frequently occur which would antecedently have been considered most improbable, and as their improbability usually arises from want of a more intimate and correct knowledge of the causes which produced them, mere improbability can rarely supply a sufficient ground for disbelieving direct and unexceptionable witnesses of the fact, where there was no room for mistake."

And again, " Mr. Hume's conclusion is highly objectionable in a philosophical point of view, inasmuch as it would leave phenomena of the most remarkable nature wholly unexplained, and would operate to the utter exclusion of all inquiry. Estoppels are odious even in judicial investigations, because they tend to exclude the truth ; in metaphysics they are intolerable. So conscious was Mr. Hume himself of the weakness of his general and sweeping position, that in the second part of his 10th section, he limits his inference in these remarkable terms, ' I beg the limitations here made may be remarked when I say that a miracle can never be proved so as to be the foundation of a system of religion ; for I own that otherwise there may *possibly* be miracles or violations of the usual course of nature of such a kind as to *admit of proof from human testimony.*"

Now this limitation, by which Hume reduced the breadth of his original proposition, is simply a too arbitrary application of a principle of criticism of testimony, in itself entirely unobjectionable, and upon which, indeed, it is one of the objects of this paper most strongly to insist. Obviously, what is regarded in the proposition thus limited is not the improbability of the fact at all, but the temptation of the witnesses to deceive, or their liability to be deceived. That is a legitimate and necessary consideration, resulting from our experience of human motives and of the effect of prepossessions, in the estimation of testimony. If the object of the witness, as of the early Christian, for example, is to persuade the world of the divine authorship of a religion, that object, and the heat and zeal with which it would probably be pursued, might undoubtedly supply a motive, proper to be taken into account, for statements of miracles performed by the author of the religion. And so the preconception of His divine powers would predispose to a facility of accepting appearances as miraculous, quite inconsistent with the cool and scientific observation which we desiderate in the witnesses. These considerations undoubtedly go to weaken the force of testimony; whether they do so in such a degree as to deprive it of all value is really a matter of individual opinion, and certainly, apart from the circumstances of each case, cannot pretend to the dignity of a universal principle of judgment. Hume has few greater admirers than myself; but I am forced to the conclusion that the celebrated *Essay on Miracles*, which he put forth with almost exulting confidence, is one of the weakest, the most ill considered, and the most inconsistent pieces of reasoning with which I am acquainted. It has been completely overthrown by three writers who have dealt with it, and of whom the later do not appear to have met with the earlier refutations; by Mr. Starkie, by Mr. Babbage, in the *Ninth Bridgwater Treatise*, and by Mr. A. R. Wallace, in the Introduction to his *Miracles and Modern Spiritualism*.

I have endeavoured to point out the fallacy of what seems to me a false application of the principle that evidence should be proportioned to probability. I will now attempt to state, in an abstract form, what I submit is its true result in our experience of testimony. If it is possible to assign a ratio of probability to a fact, not being one subject to exact computation, it is also possible to assign a similar ratio to the value of evidence, for *the value of evidence is just the probability against its being forthcoming for that which is not a fact.* If it is legitimate to consider the probability of a fact apart from the evidence for it, so it is legitimate to consider the general value of a particular quality and amount of testimony apart from the probability of any special fact to which it may be applied. No antecedent preference is due to the one probability over the other if they are equal, but the result is that precisely in proportion as both the fact is improbable and the evidence is probable, *you will not get* the evidence for the fact, that is to say, just in that proportion you are *unlikely* to get it. And if we find, and find often, evidence which we deem to be good, for a fact which we deem to be improbable, of one of two things we may be certain, either we have miscalculated the value of the evidence or the probability of the fact. Now in relation to facts new to our experience, to facts of which the proof of their possibility is also the proof of their existence, which of these alternatives is the most probable? Whatever induction experience may afford of what may be called the abstract value of evidence—that is without regard to the antecedent probabilities of the fact to be proved—is positive and affirmative. It is constantly being verified. It depends on tests and criteria, the efficiency of which are also being constantly guaranteed by experience. How stands the case with that other negative induction to which it is opposed? *The probability in its favour is just the probability that good evidence will not be forthcoming to contradict it.* It is a probability which arises entirely from the absence of evidence. It is impossible to conceive more vicious reasoning than that which would make it a ground of rejecting evidence. It depends on the proposition : "If this were true, we should have had the evidence before"—which amounts to this, as has been pointed out by Mr. Starkie, and by Mr. A. R. Wallace, in the admirable Introduction to his book, *Miracles and Modern Spiritualism*, that no new fact can ever be proved by testimony. And I cannot conclude this part of the argument better than by quoting that writer's neat dilemma in reply to Hume : "If the fact were possible, such evidence as we have been considering would prove it; if it were not possible, that evidence would not exist."

Something remains to be said on the effect of cumulative evidence. The late Mr. Babbage, in the *Ninth Bridgwater Treatise*, has worked out an elaborate mathematical refutation of Hume's principle. And he concludes that if any definite measure of improbability, however large, be adopted, that is to say, if the improbability be short of infinite (and no one has ever contended that it is this—or, in other words, that the fact is *impossible*), a miracle, so called, can be proved by testimony. Taking m as the measure of improbability, he says, "It follows, therefore, that however large m may be, however great the quantity of experience against the occurrence of a miracle (provided only that there are persons whose statements are more frequently correct than incorrect, and who give their testimony in favour of it without collusion), a certain number, n, can always be found, so that it shall be a greater improbability that their unanimous statement shall be a falsehood than that the miracle shall have occurred." Taking the case of only six witnesses who will speak the truth, and are not themselves deceived in ninety-nine cases out of a hundred, Mr. Babbage deduces the result that the improbability of their independent concurrence in testifying to what is not a fact is five times as great as an assumed improbability of two hundred thousand millions to one against the miracle which they are supposed to attest, or it is one billion to that number. And it hardly needs demonstration that the same result is arrived at by increasing the number of witnesses in proportion to any definite numerical deduction from the value of the individual testimony of each. To this scientific authority I will add a legal one to the same effect. "It would," says Mr. Starkie, in his treatise on the *Law of Evidence*, "theoretically speaking, be improper to omit to observe that the weight and force of the united testimony of numbers upon abstract mathematical principles, increases in a higher ratio than that of the mere number of such witnesses. Upon these principles, if definite degrees of probability could be assigned to the testimony of each witness, the resulting probability in favour of their united testimony would be obtained not by the mere addition of the numbers expressing the several probabilities, but by a process of multiplication." Now it is obvious that in applying these principles to a *class* of alleged facts denied on the ground of antecedent improbability, we ought to take, in computing the cumulative force of testimony, not simply the testimony which this or that fact of the class can adduce, but all the testimony which exists for all similar alleged facts comprised in the class. Let M represent the class, comprising under it *a b c d*, particular alleged instances. We may state the result in either one or two ways. Either we may oppose the improbability of (class) M to the cumulative evidence of *a b c d*, taken together; or, taking *a* by itself, we may say that the improbability against *a* is the improbability of M, the class, *minus* the probability resulting from the cumulative testimony in favour of *b, c,* and *d*, taken together. Now to apply the foregoing considerations to cases of actual occurrence. I could not go into details here without protracting this paper beyond reasonable limits, but the cases I shall take are already familiar to many in this room, and as they are on record, with the utmost particularity of description, others may be referred to the printed accounts. I select then a number of testimonies to distinct facts of the same class, namely, of physical effects produced by means unknown to science, and each depending on the introduction of new physical conditions by special human organisms, which, as before stated, and not any particular effect, is the *fact* really, if at all, opposed to experience. Let me again request you to keep this clearly in mind. If I say that an effect depends upon the powers of a certain person, your experience is evidently not opposed to the effect except so far as it is opposed to the existence of such powers in members of the human race. Your experience of the uniform course of physical nature is wholly and absolutely irrelevant. Nobody has ever asserted that these things would occur in your presence alone. If you are to bring the experience argument to bear at all, it must be in denying the alleged *conditions* of their occurrence—the chief of these conditions, in this case, relating to the personality of individuals. That premised, the several alleged facts, I take, belong to the same class—namely, those that depend on the presence of persons reputed to be *psychics*, or mediums. The first is the experiment recorded in the April number of the *Quarterly Journal of Science* of Professor Zöllner and other German scientists with Dr. Slade. In this, as in the other cases to be presently mentioned, I have taken the testimony of well-known men of scientific eminence, because, although their veracity may not be worth more than that of other witnesses to these facts, it may be called a *known quantity.* The improbability of Zöllner's *lying* would, I imagine, be admitted to exceed Mr. Babbage's 100 to 1. And so also of the others to be named. But how are we to assign a value to the improbability of his being deceived? Now here, I must remind you, the improbability of the fact attested is wholly beside the question. That is a matter to be taken into calculation subsequently. For the present purpose the probability of his being deceived or mistaken is just what it would have been if he was performing the most ordinary experiment in the world, *under the same conditions of observation, and with, of course, the same suppositions of a motive and design to deceive him.* When we have got this value, then we will set off against it the improbability of the fact. But to consider the latter at present would be just as if, having to subtract an unknown quantity, x, from a given number, say 10, we began by subtracting 10 from x, and so made the problem 10—(x−10) instead of 10-x, an algebraical begging of the question. Regarding, then, the experiment without this prejudice, I should say no numeral would be considered quite high enough to express the improbability of Zöllner's being deceived. Add to this, the improbability of his colleagues also being deceived. But whatever value we determine upon, is it to be opposed by itself to the improbability of the fact, which would then be proper to be considered? No; for look at the next case of the same class. That shall be the electrical test experiment of Mr. Crookes with Mrs. Fay, at his own house, assisted by several Fellows of the Royal Society, as well as by our president, Mr. Serjeant Cox, who all agreed in the conclusive nature of the experiment. Lying again is out of the question, practically. Deception by the medium? Inaccuracy of observation? A scientific test, devised by the most competent experts, the nature of it not explained to the medium till she, who may almost be assumed to be a scientifically ignorant young woman, is in the house (that of Mr. Crookes), the apparatus unknown to her, and its working watched and recorded from minute to minute. The results beyond all explicable power of production, even had the medium been herself an accomplished electrician, and intimately versed with the apparatus. In calculating probabilities, the same observations are applicable here as to the case of Professor Zöllner. But the improbability of deception here must be added, in the ratio pointed out by Mr. Babbage and Mr. Starkie, to that of the former case. Take yet another, and here again one at least of the witnesses is a man of high scientific standing—Lord Lindsay, who has recently been elected on the Council of the Royal Society. He describes the levitation of Mr. Daniel Home, and his floating in and out of a window seventy feet from the ground by bright moonlight. I will read the account in Lord Lindsay's own words :—

"I was sitting with Mr. Home and Lord Adare, and a cousin of his. During the sitting Mr. Home went into a trance, and in that state was carried out of the window in the room next to where we were, and was brought in at our window. The distance between the windows was about 7 feet 6 inches, and there was not the slightest foothold between them, nor was there more than a 12-inch projection to each window, which served as a ledge to put flowers on.

"We heard the window in the next room lifted up, and almost immediately after we saw Home floating in the air outside our window.

"The moon was shining full into the room ; my back was to the light, and I saw the shadow on the wall of the window-sill, and Home's feet about six inches above it. He remained in this position for a few seconds, then raised the window and glided into the room, feet foremost, and sat down.

"Lord Adare then went into the next room to look at the window from which he had been carried. It was raised about 18 inches, and he expressed his wonder how Mr. Home had been taken through so narrow an aperture.

"Home said, still entranced, 'I will show you,' *and then with his back to the window he leaned back, and was shot out of the aperture, head first, with the body rigid, and then returned quite quietly.*

"The window is about 70 feet from the ground. I very much doubt whether any skilful tight-rope dancer would like to attempt a feat of this description, where the only means of crossing would be by a perilous leap, or being borne across in such a manner as I have described, placing the question of the light aside.

"LINDSAY.

"*July 14th, 1871.*"

I will call one other witness before you, likewise of scientific position and attainments, *begging you to remember that these are only specimen cases.* It is Dr. Lockhart Robertson, one of the Visitors in Lunacy. Among other phenomena which took place in his own house, in the presence of himself and his own friends, the medium being a Mr. Squire, Dr. Robertson describes the following:—"A heavy circular chair, made of birch and strongly constructed, was lifted a somersault in the air and thrown on the bed, the left hand only of Mr. Squire being held on the surface, his other hand held, and his legs being tied to the chair on which he sat. The table was afterwards twice lifted on to the head of the writer and of Mr. Squire. . . . . At the writer's request this table was afterwards smashed and broken, and one fragment thrown across the room, the table at the time being held by the writer and Mr. Squire. This occurred in half a minute. The writer has since vainly endeavoured with all his strength to break one of the remaining legs. The one broken was rent across the grain of the wood." Dr. Robertson states that all this took place in the dark, but probably, looking at the nature of the phenomena and the conditions described, most candid persons would be of the opinion, he concludes by expressing, "that fraud was utterly and entirely impossible and impracticable." I will add just one other testimony of Lord Lindsay:—"A friend of mine was very anxious to find the will of his grandmother, who had been dead forty years, but could not even find the certificate of her death. I went with him to the Marshall's, and we had a *séance;* we sat at a table, and soon the raps came. My friend asked his questions *mentally;* he went over the alphabet himself, or sometimes I did so, not knowing the question. We were told the will had been drawn by a man named Walker, who lived in Whitechapel; the name of the street and the number of the house were given. We went to Whitechapel, found the man, and subsequently, through his aid, obtained a copy of the draft. He was quite unknown to us, and had not always lived in that locality, for he had seen better days. The medium could not possibly have known anything about the matter, and even if she had, her knowledge would have been of no avail, as all the questions were mental ones."

If you would be rational, do not laugh at these cases one by one, but study the evidence for each of them separately, and then appreciate their cumulative force, as belonging to the same class. Then, if you please, set off the improbability arising from your own and others' ignorance. I don't know if you will estimate that at Babbage's two hundred thousand millions, but if so, you are bound to show—mind, once more, without any reference, express or tacit, to the improbability of the facts—why the evidence should be estimated at less than Babbage's billion, or rather, since we have here more than six witnesses whose testimony for any ordinary fact would have so great a value, at this billion multiplied in a greater ratio than my small mathematical powers could easily calculate.

But, in fact, I place the argument far higher than either Mr. Starkie or Mr. Babbage, though I believe I am in accord with Mr. Wallace. Both the former assumed that there is an antecedent improbability to be deducted from the value of the positive testimony. I deny that altogether. I say that an improbability arising from want of evidence—which is the nature of these negative inductions—*is just the improbability that evidence will be forthcoming.* When you have got the evidence the improbability vanishes just in proportion to the value of the evidence *per se.* What you mean by the improbability of a fact beyond experience is that it is probably impossible or not *in rerum natura.* What conceivably legitimate measure of this probability can you adopt than that which also determines the relation between evidence and fact? The fallacy consists in assuming any numerical value whatever for such antecedent improbabilities apart from this relation. Say that the best single human testimony has a value of 100 to 1. Now to-day, because I have never had that evidence, I say the probability against the fact is represented in my mind as 1000 to 1. To-morrow I get the evidence of that intrinsic value of 100 to 1, and I say, "Oh, but the adverse probability is 1000 to 1, and the value of this evidence must be reduced accordingly to a minus quantity." This surely is unreasonable. But I may quite logically say, "Inasmuch as this 100 to 1 evidence has never been forthcoming, it raises in my mind a presumption worth

1000 to 1 that such evidence never will be forthcoming." If the evidence arrives after all, there is no presumption against its *truth*. We have a right to our surprise, but not to our incredulity. Because there was no evidence, we thought there was no fact. We had a right to think so. But the moment we have evidence we are in the region of evidence whose intrinsic value we have to estimate, all presumptions being henceforth merely impertinent. The case is, of course, very different when we are dealing with actual, ascertainable probabilities, as the probability of a given ball being drawn by chance from a hundred others. Then the chance being real, and not merely supposititious, we properly set it off against the evidence. But in the other case the evidence destroys the supposition, precisely in proportion to its own intrinsic value.

But even allowing the presumption to co-exist with the evidence, it has appeared that if no other evidence of similar facts had existed from the beginning of recorded time to the present besides these three cases I have mentioned, the probability in their favour would still be greater than the probability against them. You are instinctively repelled by this statement; so am I. We all feel that there must be something wrong somewhere. And so there is. It is not that the hypothesis is an impossible one. Mr. Babbage has made a very ingenious supposition. He has conceived the course of nature to be like a machine constructed on the principle of his own calculating engine. A thousand revolutions of the wheel shall bring up only square numbers, but the machine shall be constructed so that the thousand and first shall show a cube number—a "miracle." We can conceive that certainly. And so a man might be born to-day who should be the first of mankind born with these abnormal powers we have been considering. But all observed analogies protest against this supposition of a purely exceptional fact, even though we may conceive such a fact to be subsumed under a higher law of extremely rare application. If we have once proved the fact under its own conditions, it is in the highest degree propable that the law of its occurrence is in constant operation. To suppose that it is not is to encounter a new improbability, and it is this new improbability which repelled us just now in the supposition that no other similar cases had existed in human experience. We should expect to find them in every age. See now how we have shifted the onus of improbability. The proved case in the present makes such cases in the past highly probable; in other words, experience cannot have been truly opposed to that which has just been proved on the assumption that experience *is* opposed to it. And what do we find in fact? Why, that records of occult phenomena, and especially of such as occur through the mediation of particular individuals, form an appreciable part of the literature of every generation of men since the invention of printing, and anterior to that we have, besides the manuscript accounts of antiquity, the universal belief of mankind, which must presumably have rested on experience. Addison, indeed, speaks of the "general testimony of mankind" in favour of those facts to which eighteenth century scepticism—a product of intellectual causes which have been traced by Mr. Lecky—has unwarrantably opposed that very general testimony. I have said nothing of the innumerable mob of witnesses in the present time, and in almost every country in the world, to whose separate and individual testimony we are unable to assign a positive value. I have said nothing even of that respectable array of known and in various ways distinguished witnesses whom we have still among us, or who have recently deceased. I have said nothing of the admission of experts in the art of conjuring—that art to which such illimitable powers are ascribed by the credulity of the incredulous—of the celebrated conjurer Houdin, of the celebrated conjurer Bellachini, of the celebrated conjurer John Nevil Maskelyne, the latter of whom I publicly challenged in the *Examiner* newspaper to explain away, if he could, certain printed and published admissions of his own to the existence of phenomena of this class not produced by trickery.* I am not attempting the prodigious task of estimating in figures the cumulative evidence for the phenomena called spiritualistic, a Pelion piled upon an Ossa of testimony, and which would crush any logical resistance, but not the illogical power of that against which, it is said, the very gods strive vainly. I charge this stupidity with gross ignorance of the principles upon which evidence should be estimated; and I have traced this ignorance to four fallacies ; First, to the confusion of the positive affirmative induction which we legitimately draw of the course of nature under ordinary conditions of observation, with the negative induction from inexperience, of the non-existence of other conditions. Secondly, to the assumption that this inexperience, in fact, exists, as the ground even of this negative, far more limited, and far less valid induction, an assumption which is made by an arbitrary rejection of historic evidence. Thirdly, to the assumption that antecedent improbability thus arising *can* co-exist with testimony of a certain assignable value. Fourthly, to neglecting to estimate the *cumulative* force of testimony.

That these fallacies are, nevertheless, sanctioned by

* Notice of the terms of the above reference to Mr. Maskelyne was sent to the latter, with a card of admission to hear the paper read, available for Maskelyne himself, or for any friend by whom he might wish to be represented, and who might make any statement by permission of the chairman. For Mr. Maskelyne's admissions see Appendix C.

common consent, and by authority, need not surprise us. It is a popular error that priests have been the greatest enemies to science. It has been the "common sense" of each generation, supported and sanctioned by the highest scientific authorities of the day, that has always been found opposed to the reception of evidence conflicting with presumptions which have their origin in ignorance. It was not a Churchman, but a very learned professor, notorious for his anti-religious tendencies, who refused to look through Galileo's telescope. Was it religious persecution or popular and scientific ridicule that Harvey, Jenner, Franklin, Young, Stephenson, Arago, and Gregory, encountered for their respective discoveries and ideas? It is significant that in an American book, called the *Warfare of Science*, that was republished in England last year under the avowed patronage of Professor Tyndall, there is much that is well and eloquently told of the wrongs of science at the hands of religious bigotry, but not one word of the constant and determined obstruction of scientific men.

To avoid misapprehension I wish to add one remark. In speaking of the abstract value of testimony I have not for a moment meant to imply that testimony, or evidence generally, can be appreciated without reference to the nature of the fact attested. It is only the assumed improbability of the fact which I have regarded as a separable factor. But in accounts of the extraordinary there are undoubtedly elements of fallacy which only a very inexperienced judge of testimony would ignore. For instance, we may almost appropriate a special set of motives to such narrations. The mere vanity of producing an impression of wonder, or of making out an unanswerable case, is responsible for many a false or highly coloured account. There is the temptation to support a hasty exaggeration by a specific falsehood, or by suppression of truth. Then, again, the fact may be of such a nature that the whole value of the testimony depends on minuteness and accuracy of observation. Regard to time prevents my doing more than advert to these considerations. Only to each case as it arises can their proper weight be assigned. Unfortunately there is assigned to them an enormously exaggerated weight in general, without reference to particular cases at all, and this because it is assumed to be more probable that the evidence is thus vitiated than that the facts attested are true. No doubt the presumption that evidence is not good is a far more rational presumption than that evidence, however good, is false. And, moreover, it is one which can be brought to the test of examination, whereas the latter cannot. We can show whether evidence does or does not come up to a certain standard, and if it does, the presumption is falsified; but to the man who says, " I won't listen, and I dont care how good your evidence may be," we can have nothing further to say.

In conclusion, I will lay down the following proposition broadly. A negative probability, by which I mean an inference of non-existence from the absence of evidence, cannot in the least affect the value of positive evidence of existence. It is only provisional. It vanishes at the touch of sufficient evidence; and sufficient evidence I define, for this purpose, to be evidence which would establish a fact — having strict regard to the nature of the fact—as to which there was no antecedent presumption or probability for or against. Would I therefore accept the statement of a casual stranger as to some unheard-of marvel with the same facility that I would accept his statement as to its having rained somewhere yesterday—a fact which may be said to answer the description of having no antecedent presumption either way? Certainly not, for I have said that the nature of the fact is to be regarded, not as probable or improbable, but as communicating elements of fallacy to testimony. Thus understood, I say that the evidence is our whole concern, and that if it stood every test and every criticism which experience could suggest, I would accept on the strength of it any marvel in the *Arabian Nights*, or *Gulliver's Travels*. And I submit that the man who would not is the creature of prejudice and the victim of prepossessions.

APPENDIX B.

EVIDENCE OF SAMUEL BELLACHINI, COURT CONJURER AT BERLIN.

The following is a translated copy of an official document :—

No. 482 Notary's Register for 1877, drawn at Berlin, the 6th day of December one thousand eight hundred and seventy-seven, in presence of the undersigned notary, residing at Tauben-strasse, No. 42, in the jurisdiction of the Royal Supreme Court of judicature, Gustav Haagen, Counsellor, and in presence of the undersigned witnesses, personally known to the notary, of full age, who can read and write, and are residents here.

Carl Trümper, *Letter Carrier*,
Gustav Grüntz, *Letter Carrier*,
who as well as the notary, as notary and witnesses both hereby declare they have no connection wit the case, which according to pages five to nine of the Act of July the eleventh, eighteen hundred and forty-five, would exclude them from participating in this document,

Did appear this day personally before the undersigned notary, known to him and found duly qualified to act.

The prestidigitator and Court Conjurer to his Majesty the King and Emperor William I Mr.

Samuel Bellachini, residing at Grossbaaron-strasse, No. 14, which gentleman did prefer the following statement, under date Berlin, the 6th of December, in this year, and that he certified,

That the signature of my name, hereby appended, was written by me in due form I hereby acknowledge. Read, approved, and executed.

(Signed)   SAMUEL BELLACHINI.

We, the notary and witnesses, attest that the above transaction took place as herein stated; that it was in the presence of us, notary and witnesses, read aloud to the person concerned, approved by him, and signed by his own hand.

(Signed)   GUSTAV GRUNTZ,
KARL TRUMPER,
GUSTAV HAAGEN, *Notary*.

Executed at Berlin on the sixth of December, one thousand, eight hundred, and seventy-seven, and entered in the Notary's Register under the number four hundred and eighty two, for the year eighteen hundred and seventy-seven. Signed and officially stamped.

GUSTAV HAAGEN, *Counsellor and Notary*.

I hereby declare it to be a rash action to give decisive judgment upon the objective medial performance of the American medium, Mr. Henry Slade, after only one sitting, and the observations so made.

After I had, at the wish of several highly-esteemed gentlemen of rank and position, and also for my own interest, tested the physical mediumship of Mr. Slade in a series of sittings by full daylight, as well as in the evening, in his bedroom, I must, for the sake of truth, hereby certify that the phenomenal occurrences with Mr. Slade have been thoroughly examined by me with the minutest observation and investigation of his surroundings, including the table, and that I have *not in the smallest degree* found anything to be produced by means of prestidigitative manifestations, or by mechanical apparatus; and that any explanation of the experiments which took place *under the circumstances and conditions then obtaining* by any reference to prestidigitation, *to be absolutely impossible.*

It must rest with such men of science as Crookes and Wallace, in London ; Perty, in Berne ; Butlerof, in St. Petersburg ; to search for the explanation of this phenomenal power, and to prove its reality. I declare, moreover, the published opinions of laymen, as to the "How" of this subject to be premature, and according to *my* view and experience, false and one-sided. This, my declaration, is signed and executed before a notary and witnesses.

(Signed)   SAMUEL BELLACHINI.

BERLIN, 6*th December*, 1877.

APPENDIX C.

ADMISSIONS BY JOHN NEVIL MASKELYNE, AND OTHER PROFESSIONAL CONJURERS.

Mr. John Nevil Maskelyne, the well-known conjurer of the Egyptian Hall, Piccadilly, who without having been present at a single sitting with Slade, was irregularly admitted as a witness against him at Bow Street, had long been adding to the attraction of his performances by holding them out to be exposures of spiritualistic phenomena. In June and July 1873 a correspondence took place between this gentleman and a Spiritualist, in which the latter offered Mr. Maskelyne £1000 if he could reproduce certain mediumistic phenomena with the conditions under which they had been observed by three persons, one only to be a Spiritualist. The negotiation came to nothing, but the correspondence was printed, and the following extracts are quoted with a view to show that Mr. Maskelyne has himself made distinct admissions of the reality of some of such phenomena, *not due to trickery*, that he even avows them as part of his own public exhibitions, and that he merely protests that spirits of the dead have nothing to do with them. That this is not the true issue every one but Mr. Maskelyne will admit. That issue is simply *trickery by the medium, or not*. If not, the phenomena must be entitled to admission and investigation, and must give rise to scientific questions of the utmost moment.

The extracts are taken from the printed correspondence entitled as follows :—

"£1000 Reward.
Maskelyne and Cooke,
An Exposé, &c.,
By Iota.
(Proofs corrected by Mr. Maskelyne).
London, J. Burns, 15 Southampton Row, W. C."

In order to make them intelligible it should be premised that "the manifestations stated in the report of the Dialectical Society," were distinctly mediumistic, the committee of that society which made the report having been appointed for the express purpose of investigating and reporting upon spiritualistic phenomena. The report with the evidence at length is published.

On the 1st July 1873, Mr. Maskelyne, in the course of a letter to his correspondent, writes as follows:—

". . . . In accepting this challenge, I wish you distinctly to understand that I do not presume to prove that such manifestations as those stated in the report of the Dialectical Society are produced by trickery—I have never denied that such manifestations are genuine, but I contend that in them there is not one iota of evidence which proves that departed spirits have no better occupation than lifting furniture about". . . .

Agreed, Mr. Maskelyne; those Spiritualists, if any, who are not entirely of your opinion on this point, seem to deserve your imputation of credulity in the highest degree. Accordingly, the other party to the correspondence replies on the following day—

". . . . I do not care to dispute your contention

about the occupation of departed spirits. What I understand by medium-power is something which is neither mechanics, nor conjuring, nor chemistry, nor electricity, nor magnetism, nor even mesmerism, nor a combination of all or any of these, nor anything to be explained by any of the *commonly* known 'laws of nature,' and without which I defy you to equal, or even to approach, the 'so-called spiritual manifestations.'"

Then on the 6th Mr. Maskelyne again writes—" I have never stated that you cannot produce some phenomena in a genuine manner; I have done this or assisted in doing it myself, and tell my audience so at every performance; yet I am not a medium, but I know that, if I were scoundrel enough, I could soon become one, and should have no difficulty in humbugging Spiritualists to an alarming extent." Here, again, Mr. Maskelyne appears to be speaking merely of an explanation which he holds to be false, and which he believes that professed mediums must know to be false. But in urging these phenomena upon public attention we have nothing to do with spiritualistic explanations, true or false; it is the *fact* only that is in question. What Mr. Maskelyne means by saying that he tells his audience at every performance that he produces or assists in producing phenomena "in a genuine manner" (by which, as will be seen, he excludes the notion of trickery) is very doubtful. The writer has attended the performances at the Egyptian Hall frequently, but with the exception of some words at the conclusion of the cabinet *séance* which could convey no meaning to an *inexperienced* audience, Mr. Maskelyne certainly said nothing to which his above statement could apply. These occasions, however, were of much later date than the correspondence.

His correspondent replies on July 8th—

"You say you tell your audience at every performance that you admit that we have some genuine phenomena. I confess that I have never been able to understand distinctly your remarks on this head. You seem to me to say that most of the so-called phenomena are humbug, but some few genuine; that the genuine ones are produced by trickery, exactly as your own stage performance is. Nor can I gather any more from the admissions in your letters."

In a postscript to his next letter, Mr. Maskelyne says, in reference to the above, "How genuine phenomena can be produced by trickery I am at a loss to know. If you understand me thus, my remarks must be a contradiction, and I must look to them."

Robert Houdin, the great French conjurer, investigated the subject of clairvoyance with the sensitive, Alexis Didier. In the result he unreservedly admitted that what he had observed was wholly beyond the resources of his art to explain. See "Psychische Studien" for January 1878, p. 43.

"Licht, mehr Licht," a German paper published in Paris, in its number of 16th May 1880, contains a letter from the well-known professional conjurer, Jacobs, to the Psychological Society in Paris, avowing himself a Spiritualist, and offering suggestions for the discrimination of genuine from spurious manifestations.

APPENDIX D.

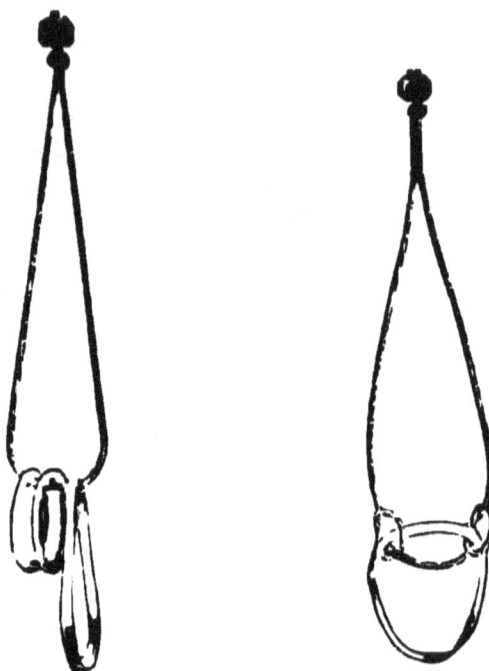

PLATE X.

STANDARD BOOKS

ON

SPIRITUALISM, MESMERISM, PSYCHOLOGY, ANTHROPOLOGY,

AND KINDRED SUBJECTS,

PUBLISHED BY W. H. HARRISON,

33 MUSEUM STREET, LONDON, W.C.

Lists of the Books are Advertised in every Number of the "Spiritualist" Newspaper.

Mr. W. H. Harrison's Publications may be obtained from Mr. W. H. Terry, 84 Russell Street South, Melbourne, Australia.

THE "SPIRITUALIST" NEWSPAPER:

A RECORD OF THE PROGRESS OF THE SCIENCE AND ETHICS OF SPIRITUALISM.

PUBLISHED WEEKLY, PRICE TWOPENCE.

Established in 1869.

The *Spiritualist*, published weekly, is the oldest Newspaper connected with the movement in the United Kingdom, and is the recognised organ of educated Spiritualists in all the English-speaking countries throughout the Globe. It also has an influential body of readers on the Continent of Europe.

The Contributors to its pages comprise the leading and more experienced Spiritualists, including many eminent in the ranks of Literature, Art, Science, and the Nobility of Europe. Among those who have published their names in connection with their communications in its columns are — His Imperial Highness Nicholas of Russia, Duke of Leuchtenberg; Prince Emile de Sayn Wittgenstein (Wiesbaden); the Lord Lindsay; the Count de Bullet; the Right Hon. the Countess of Caithness; the Hon. J. L. O'Sullivan, formerly American Minister at the Court of Portugal; the Baroness Von Vay (Austria); M. Adelberth de Bourbon, First Lieutenant of the Dutch Guard to H.M. the King of the Netherlands; the Hon. Robert Dale Owen, formerly American Minister at the Court of Naples; M. L. F. Clavairoz (Leon Favre), Consul-General of France at Trieste; the Hon. Alexandre Aksakof, St. Petersburg; Baron Von Dirckinck-Holm-

[ADVERTISEMENTS.]
W. H. HARRISON'S PUBLICATIONS.

feld (Holstein); Sir Charles Isham, Bart.; William Crookes, Esq., F.R.S., editor of *The Quarterly Journal of Science*; Captain R. F. Burton, F.R.G.S. (Discoverer of Lake Tanganyika); C. F. Varley, Esq., C.E., F.R.S.; Alfred Russel Wallace, Esq., F.R.G.S.; Miss Florence Marryat; C. C. Massey, Esq.; St. George W. Stock, Esq., M.A. (Oxon); Mr. Serjeant Cox, President of the Psychological Society of Great Britain; J. M. Gully, Esq., M.D.; Alexander Calder, Esq., President of the British National Association of Spiritualists; Epes Sargent, Esq.; Colonel H. S. Olcott, President of the Theosophical Society of New York; Dr. George Wyld; Mrs. Makdougall Gregory; W. Lindesay Richardson, Esq., M.D., Melbourne; Gerald Massey, Esq.; J. C. Luxmoore, Esq., J.P.; Mrs. Weldon (Miss Treherne); C. Carter Blake, Esq., Doc. Sci., Lecturer on Comparative Anatomy at Westminster Hospital; S. C. Hall, Esq., F.S.A.; Hensleigh Wedgwood, Esq., J.P.; Mrs. S. C. Hall; H. M. Dunphy, Esq.; Eugene Crowell, Esq., M.D., New York; Algernon Joy, Esq., M. Inst. C.E.; Stanhope T. Speer, Esq., M.D., Edinburgh; Desmond FitzGerald, Esq., M.S. Tel. E.; Robert S. Wyld, Esq., LL.D.; J. A. Campbell, Esq.; Captain John James; D. H. Wilson, Esq., M.A., LL.M. (Cantab.); the Rev. C. Maurice Davies, D.D., author of *Unorthodox London*; T. P. Barkas, Esq., F.G.S.; H. D. Jencken, Esq., M.R.I.; J. N. T. Martheze, Esq.; Charles Blackburn, Esq.; Mrs. Showers; Miss Kislingbury; William Newton, Esq., F.R.G.S.; John E. Purdon, Esq., M.B., India; H. G. Atkinson, Esq., F.G.S., author of *Letters to Miss Martineau*; and William White, Esq., author of *The Life of Swedenborg*.

Annual Subscription to Residents in the United Kingdom, 10s. 10d.; in the United States and Australia, 13s., post-free.

The *Spiritualist* is regularly on sale at the following places:—
LONDON—11 Ave Maria Lane, St. Paul's Churchyard, E.C.; PARIS—Kiosque, 246 Boulevard des Capucines, and 5 Rue Neuve des Petits Champs, Palais Royal; LEIPZIG—2 Lindenstrasse; FLORENCE—Signor G. Parisi, Via della Maltonaia; ROME—Signor Bocca, Libraio, Via del Corso; NAPLES—British Reading Rooms, 267 Riviera de Chiaja, opposite the Villa Nazionale; LIEGE—37 Rue Florimont; BUDA-PESTH—Josefstaadt Erzherzog, 23 Alexander Gasse; MELBOURNE—84 Russell Street South; SHANGHAI—Messrs. Kelly & Co.; NEW YORK—51 East Twelfth Street; BOSTON, U.S.—*Banner of Light* Office, 9 Montgomery Place; CHICAGO—*Religio-Philosophical Journal* Office; SAN FRANCISCO—319 Kearney Street; PHILADELPHIA—325 North Ninth Street; and WASHINGTON—1010 Seventh Street.

All communications on the business of the *Spiritualist* should be addressed to W. H. HARRISON, *Spiritualist* Newspaper Branch Office, 33 Museum Street, London, W.C.

[ADVERTISEMENTS.]
W. H. HARRISON'S PUBLICATIONS.

MESMERISM AND ITS PHENOMENA;
OR, ANIMAL MAGNETISM.

By the Late WM. GREGORY, M.D., F.R.S.E., Professor of Chemistry at Edinburgh University.

DEDICATED BY THE AUTHOR BY PERMISSION TO HIS GRACE GEORGE DOUGLAS-CAMPBELL, DUKE OF ARGYLL.

This second and slightly revised and abridged Edition is for its quality and size one of the Cheapest Large Works ever Published in England in connection with Spiritualism.

THE CHIEF STANDARD WORK ON MESMERISM.

Price 5s., or 5s. 6d. post-free.

CONTENTS.

CHAPTER I.—First Effects Produced by Mesmerism—Sensations—Process for Causing Mesmeric Sleep—The Sleep or Mesmeric State—It Occurs Spontaneously in Sleep-Walkers—Phenomena of the Sleep — Divided Consciousness — Senses Affected—Insensibility to Pain.

CHAPTER II.—Control Exercised by the Operator over the Subject in Various Ways—Striking Expression of Feelings in the Look and Gesture—Effect of Music—Truthfulness of the Sleeper—Various Degrees of Susceptibility—Sleep Caused by Silent Will, and at a Distance—Attraction towards the Operator—Effect in the Waking State of Commands given in the Sleep.

CHAPTER III.—Sympathy—Community of Sensations; of Emotions—Danger of Rash Experiments—Public Exhibitions of Doubtful Advantage—Sympathy with the Bystanders—Thought-Reading—Sources of Error—Medical Intuition—Sympathetic Warnings—Sympathies and Antipathies—Existence of a Peculiar Force or Influence.

CHAPTER IV.—Direct Clairvoyance or Lucid Vision, without the Eyes—Vision of Near Objects: through Opaque Bodies: at a Distance—Sympathy and Clairvoyance in regard to Absent Persons—Retrovision—Introvision.

CHAPTER V.—Lucid Prevision—Duration of Sleep, &c., Predicted—Prediction of Changes in the Health or State of the Seer—Prediction of Accidents, and of Events Affecting Others—Spontaneous Clairvoyance—Striking Case of it—Spontaneous Retrovision and Prevision—Peculiarities of Speech and of Consciousness in Mesmerised Persons—Transference of Senses and of Pain.

[ADVERTISEMENTS.]
W. H. HARRISON'S PUBLICATIONS.

CHAPTER VI.—Mesmerism, Electro-Biology, Electro-Psychology and Hypnotism, essentially the same—Phenomena of Suggestions in the Conscious or Waking State—Dr. Darling's Method and its Effects—Mr. Lewis's Method and its Results—The Impressible State—Control Exercised by the Operator—Gazing - Mr Braid's Hypnotism—The Author's Experience—Importance of Perseverance—The Subject must be Studied.

CHAPTER VII.—Trance, Natural and Accidental; Mesmeric—Trance Produced at Will by the Subjects—Colonel Townsend—Fakeers—Extasis—Extatics not all Impostors—Luminous Emanations—Extasis often Predicted—M. Cahagnet's Extatics—Visions of the Spiritual World.

CHAPTER VIII.—Phreno-Mesmerism—Progress of Phrenology—Effects of Touching the Head in the Sleep—Variety in the Phenomena—Suggestion—Sympathy—There are Cases in which these Act, and others in which they do not Act—Phenomena Described—The Lower Animals Susceptible of Mesmerism—Fascination among Animals—Instinct—Sympathy of Animals—Snail Telegraph Founded on it.

CHAPTER IX.—Action of Magnets, Crystals, &c., on the Human Frame—Researches of Reichenbach—His Odyle is Identical with the Mesmeric Fluid of Mesmer, or with the Influence which Causes the Mesmeric Phenomena—Odylic or Mesmeric Light—Aurora Borealis Artificially Produced—Mesmerised Water—Useful Applications of Mesmerism—Physiological, Therapeutical, &c.—Treatment of Insanity, Magic, Divination, Witchcraft, &c., explained by Mesmerism, and Traced to Natural Causes—Apparitions—Second Sight is Waking Clairvoyance—Predictions of Various Kinds.

CHAPTER X.—An Explanation of the Phenomena Attempted or Suggested—A Force (Odyle) Universally Diffused, Certainly Exists, and is Probably the Medium of Sympathy and Lucid Vision—Its Characters—Difficulties of the Subject—Effects of Odyle—Somnambulism—Suggestion—Sympathy—Thought-Reading—Lucid Vision—Odylic Emanations—Odylic Traces followed up by Lucid Subjects—Magic and Witchcraft—The Magic Crystal, and Mirror, &c., Induce Walking Clairvoyance—Universal Sympathy—Lucid Perception of the Future.

CHAPTER XI.—Interest felt in Mesmerism by Men of Science—Due Limits of Scientific Caution—Practical Hints—Conditions of Success in Experiments—Cause of Failure—Mesmerism a Serious Thing—Cautions to the Student—Opposition to be Expected.

CHAPTER XII.—Phenomena Observed in the Conscious or Waking State—Effects of Suggestion on Persons in an Impressible State—Mr. Lewis's Experiments with and without Suggestion—Cases—Dr. Darling's Experiments—Cases—Conscious or Waking Clairvoyance, Produced by Passes, or by

[ADVERTISEMENTS.]
W. H. HARRISON'S PUBLICATIONS.

Concentration—Major Buckley's Method—Cases—The Magic Crystal Induces Waking Lucidity when Gazed at—Cases—Magic Mirror—Mesmerised Water—Egyptian Magic.

CHAPTER XIII.—Production of the Mesmeric Sleep—Cases—Eight out of Nine Persons Recently Tried by the Author Thrown into Mesmeric Sleep—Sleep Produced without the Knowledge of the Subject—Suggestion in the Sleep—Phreno-Mesmerism in the Sleep—Sympathetic Clairvoyance in the Sleep—Cases—Perception of Time—Cases; Sir J. Franklin; Major Buckley's Case of Retrovision.

CHAPTER XIV.— Direct Clairvoyance — Cases — Travelling Clairvoyance—Cases—Singular Visions of Mr. D.—Letters of Two Clergymen, with Cases—Clairvoyance of Alexis—Other Cases.

CHAPTER XV.—Trance—Extasis—Cases—Spontaneous Mesmeric Phenomena—Apparitions—Predictions.

CHAPTER XVI.—Curative Agency of Mesmerism—Concluding Remarks, and Summary.

SPIRIT-PEOPLE:
A Scientifically Accurate Description of Manifestations Recently Produced by Spirits,

AND

SIMULTANEOUSLY WITNESSED BY THE AUTHOR AND OTHER OBSERVERS IN LONDON.

BY WILLIAM H. HARRISON.

Limp cloth, red edges, price 1s.; post-free, 1s. 1d.

OPINIONS OF THE PRESS.

"As a dispassionate scientific man, he appears to have investigated the subject without preconceived ideas, and the result of his examination has been to identify his opinions with those of Messrs. Varley, Crookes, and Wallace, in favour not only of the absolute reality of the phenomena, but also of the genuineness of the communications alleged to be given by the spirits of the departed."—*Public Opinion.*

"At the outset of his booklet Mr. Harrison disclaims any intention of proselytising or forcing his opinion down non-Spiritualistic throats, and it is only fair to admit that the succeeding pages are remarkably free from argument and deduction, albeit bristling with assertions of the most dumbfounding nature."—London *Figaro.*

"He neither theorises nor dogmatises, nor attempts to make converts to his views. He states occurrences and events, or what he believes did really happen, in a remarkably clear and narrative style, without any attempt at advocacy or argument."—*South Wales Daily News.*

[ADVERTISEMENTS.]
W. H. HARRISON'S PUBLICATIONS.

Price 5s., crown 8vo, richly gilt.

THE LAZY LAYS,
AND PROSE IMAGININGS.

By WILLIAM H. HARRISON.

An Elegant and Amusing Gift-Book of Poetical and Prose Writings, Grave and Gay.

The Gilt Device on the Cover designed by FLORENCE CLAXTON and the AUTHOR.

CONTENTS.

PART I.—*Miscellaneous Poems and Prose Writings*.

1. The Lay of the Lazy Author.—2. The Song of the Newspaper Editor.—3. The Song of the Pawnbroker.—4. The Castle. —5. The Lay of the Fat Man.—6. The Poetry of Science.—7. How Hadji Al Shacabac was Photographed (a letter from Hadji Al Shacabac, a gentleman who visited London on business connected with a Turkish Loan, to Ali Mustapha Ben Buckram, Chief of the College of Howling Dervishes at Constantinople).— 8. The Lay of the Broad-Brimmed Hat.—9. St. Bride's Bay.— 10. The Lay of the Market Gardener.—11. "Fast Falls the Eventide."—12. Our Raven.—13. Materialistic Religion.—14. The Lay of the Photographer.—15. How to Double the Utility of the Printing Press.—16. The Song of the Mother-in-Law.—17. *Wirbel-bewegung*.—18. "Poor Old Joe!"—19. The Human Hive.—20. The Lay of the Mace-Bearers.—21. A Love Song.— 22. A Vision.—23. "Under the Lines."—24. The Angel of Silence.

PART II.—*The Wobblejaw Ballads, by Anthony Wobblejaws.*

25. The Public Analyst.—26. General Grant's Reception at Folkestone.—27. The Rifle Corps.—28. Tony's Lament.—29. The July Bug.—30. The Converted Carman.

From the *Graphic*.

"Those who can appreciate genuine, unforced humour should not fail to read 'The Lazy Lays and Prose Imaginings.' Written, printed, published, and reviewed by William H. Harrison (38 Great Russell Street). Both the verses and the short essays are really funny, and in some of the latter there is a vein of genial satire which adds piquancy to the fun."

[ADVERTISEMENTS.]
W. H. HARRISON'S PUBLICATIONS.

Price 5s., crown 8vo, cloth, red edges.
PSYCHOGRAPHY.
BY M.A., OXON.

A Work dealing with the Psychic or Spiritual Phenomenon of the production of written messages without mortal hands. Full of well-authenticated examples.

SYNOPSIS OF CONTENTS.
LIST OF WORKS BEARING ON THE SUBJECT.
PREFACE.
INTRODUCTION.
PSYCHOGRAPHY IN THE PAST: Guldenstubbé—Crookes.
PERSONAL EXPERIENCES IN PRIVATE, AND WITH PUBLIC PSYCHICS.
    GENERAL CORROBORATIVE EVIDENCE.
I. That attested by the Senses—
    1. *Of Sight.*
        Evidence of Mr. E. T. Bennett.
        ,,   a Malvern Reporter.
        ,,   Mr. James Burns.
        ,,   Mr. H. D. Jencken.
    2. *Of Hearing.*
        Evidence of Mr. Serjeant Cox.
        ,,   Mr. George King.
        ,,   Mr. Hensleigh Wedgwood.
        ,,   Miss * * * *
        ,,   Canon Mouls.
        ,,   Baroness Von Vay.
        ,,   G. H. Adshead.
        ,,   W. P. Adshead.
        ,,   E. H. Valter.
        ,,   J. L. O'Sullivan.
        ,,   Epes Sargent.
        ,,   James O. Sargent.
        ,,   John Wetherbee.
        ,,   H. B. Storer.
        ,,   C. A. Greenleaf.
        ,,   Public Committee with Watkins.
II. From the Writing of Languages Unknown to the Psychic.
    Ancient Greek—Evidence of Hon. R. Dale Owen and Mr. Blackburn. (Slade.)
    Dutch, German, French, Spanish, Portuguese. (Slade.)
    Russian—Evidence of Madame Blavatsky. (Watkins.)
    Romaic—Evidence of T. T. Timayenis. (Watkins.)
    Chinese. (Watkins.)

[ADVERTISEMENTS.]
W. H. HARRISON'S PUBLICATIONS.

III. From Special Tests which Preclude Previous Preparation of the Writing.
Psychics and Conjurers Contrasted.
Slade before the Research Committee of the British National Association of Spiritualists.
Slade Tested by C. Carter Blake, Doc. Sci.
Evidence of Rev. J. Page Hopps. (Slade.)
,, W. H. Harrison. (Slade.)
,, J. Seaman. (Slade.)
Writing within Slates securely screwed together.
Evidence of Mrs. Andrews and J. Mould.
Dictation of Words at the Time of the Experiment.
Evidence of A. R. Wallace, F.R.G.S.
,, Hensleigh Wedgwood, J.P.
,, Rev. Thomas Colley.
,, W. Oxley.
,, George Wyld, M.D.
,, Miss Kislingbury.
Writing in Answer to Questions Inside a Closed Box.
Evidence of Messrs. Adshead.
Statement of Circumstances under which Experiments with F. W. Monck were conducted at Keighley.
Writing on Glass Coated with White Paint.
Evidence of Benjamin Coleman.

Letters Addressed to the *Times* on the Subject of the Prosecution of Henry Slade by Messrs. Joy, Joad, and Professor Barrett, F.R.S.E.
Evidence of W. H. Harrison, Editor of the *Spiritualist*.

SUMMARY OF FACTS NARRATED.

DEDUCTIONS, EXPLANATIONS, AND THEORIES.

The Nature of the Force : Its Mode of Operation.
Evidence of C. Carter Blake, Doc. Sci., and Conrad Cooke, C.E.
Detonating Noises in Connection with it.
Evidence of Hensleigh Wedgwood, J. Page Hopps, Thomas Colley.
Method of Direction of the Force.
Dr. Collyer's Theory.
Dr. George Wyld's Theory.
The Oculist's Theory.
The Spiritualist's Theory.

APPENDIX.
The Court Conjurer of Berlin on Slade.
Slade with the Grand Duke Constantine.
Recent Experiment with Monck.

[ADVERTISEMENTS.]
W. H. HARRISON'S PUBLICATIONS.

Price 5s., richly gilt, cloth ; or 3s. 6d., red edges, cloth,
imperial 8vo.

"RIFTS IN THE VEIL."

A Collection of Choice Poems and Prose Essays given through Mediumship, also of Articles and Poems written by Spiritualists. A useful book to place in Public Libraries, and to present or lend to those who are unacquainted with Spiritualism.

The 3s. 6d. Edition consists of one of the cheapest works of very high quality ever published in connection with Spiritualism; the book contains much about the religious aspects of the subject, and its relation to Christianity.

CONTENTS.

1. Introduction : The Philosophy of Inspiration.
2. "O ! Beautiful White Mother Death." Given through the Trance-Mediumship of Cora L. V. Tappan-Richmond.
3. The Apparition of Sengireef. By Sophie Aksakof.
4. The Translation of Shelley to the Higher Life. Given through the Trance-Mediumship of T. L. Harris.
5. Gone Home. Given through the Trance-Mediumship of Lizzie Doten.
6. The Birth of the Spirit. Given through the Trance-Mediumship of Cora L. V. Tappan-Richmond.
7. Angel Guarded.
8. An Alleged Post-Mortem Work by Charles Dickens. How the writings were produced ; The Magnificent Egotist, Sapsea ; Mr. Stollop Reveals a Secret ; A Majestic Mind Severely Tried ; Dwellers in Cloisterham ; Mr. Peter Peckcraft and Miss Keep ; Critical Comments.
9. The Spider of the Period. By Georgina Weldon (Miss Treherne) and Mrs. ———.
10. Margery Miller. Given through the Trance-Mediumship of Lizzie Doten.
11. Ode by " Adamanta."
12. Swedenborg on Men and Women. By William White, author of " The Life of Swedenborg."
13. Resurgam. By Caroline A. Burke.

[ADVERTISEMENTS.]
W. H. HARRISON'S PUBLICATIONS.

14. Abnormal Spectres of Wolves, Dogs, and other Animals. By Emile, Prince of Wittgenstein.
15. To You who Loved Me. By Florence Marryat.
16. Desolation. By Caroline A. Burke.
17. Truth. Given through the Mediumship of "M.A., Oxon."
18. Thy Love. By Florence Marryat.
19. Haunting Spirits. By the Baroness Adelma Von Vay (Countess Wurmbrand).
20. Fashionable Grief for the Departed.
21. The Brown Lady of Rainham. By Lucia C. Stone.
22. A Vision of Death. By Caroline A. Burke.
23. A Story of a Haunted House. By F. J. Theobald.
24. "Love the Truth and Peace." By the Rev. C. Maurice Davies, D.D.
25. The Ends, Aims, and Uses of Modern Spiritualism. By Louisa Lowe.
26. De Profundis. By Anna Blackwell.
27. Ancient Thought and Modern Spiritualism. By C. Carter Blake, Doc. Sci., Lecturer on Comparative Anatomy at Westminster Hospital.
28. Die Sehnsucht. Translated by Emily Kislingbury, from the German of Schiller.
29. The Relation of Spiritualism to Orthodox Christianity. Given through the Mediumship of "M.A., Oxon."
30. A *Séance* in the Sunshine. By the Rev. C. Maurice Davies, D.D.
31. "My Saint." By Florence Marryat.
32. The Deathbeds of Spiritualists. By Epes Sargent.
33. The Touch of a Vanished Hand. By the Rev. C. Maurice Davies, D.D.
34. Death. By Caroline A. Burke.
35. The Spirit Creed. Through the Mediumship of "M.A., Oxon."
36. The Angel of Silence. By W. H. Harrison.
37. The Prediction. By Alice Worthington (Ennesfallen).
38. Longfellow's Position in Relation to Spiritualism.
39. Spiritual Manifestations among the Fakirs in India. By Dr. Maximilian Perty, Professor of Natural Philosophy, Berne; Translated from "Psychic Studies" (Leipzig), by Emily Kislingbury.
40. The Poetry of Science. By W. H. Harrison.
41. Meditation and the Voice of Conscience. By Alex. Calder.
42. Dirge. By Mrs. Eric Baker.
43. Epigrams. By Gerald Massey.
44. Some of the Difficulties of the Clergy in Relation to Spiritualism. By Lisette Makdougall Gregory.
45. Immortality. By Alfred Russel Wallace, F.R.G.S.
46. A Child's Prayer. By Gerald Massey.

[ADVERTISEMENTS.]
W. H. HARRISON'S PUBLICATIONS.

Price 5s., cloth, crown 8vo, red edges.

SPIRIT-IDENTITY.

BY M.A., OXON.

Contains strong evidence that some of the Spirits who communicate through mediumship are the departed individuals they say they are.

SYLLABUS OF CONTENTS.

INTRODUCTION.
Difficulties in the way of the investigation.
Divergent results of investigators.
Attitude of public opinion represses publication.
This results also from the nature of the facts themselves.
The Intelligent Operator has to be reckoned with.
The investigator has little choice in the matter.
The higher phenomena are not susceptible of demonstration by the scientific method.
The gates being ajar, a motley crowd enters in.
We supply the material out of which this is composed.
No necessity to have recourse to the diabolic element.
Neglect of conditions proper for the investigation.
Agencies other than those of the departed.
Sub-human spirits—the liberated spirit of the psychic.
These have had far more attributed to them than they can rightly claim.
Specialism in Spiritualism.
Religious aspects of the question.
Notes of the age.
The place of Spiritualism in modern thought.

THE INTELLIGENT OPERATOR AT THE OTHER END OF THE LINE.
Scope of the inquiry.
The nature of the Intelligence.
What is the Intelligence?

[ADVERTISEMENTS.]
W. H. HARRISON'S PUBLICATIONS.

Difficulties in the way of accepting the story told by the Intelligence.
Assumption of great names.
Absence of precise statement.
Contradictory and absurd messages.
Conditions under which good evidence is obtained.
Value of corroborative testimony.
Personal experiences—
Eleven cases occurring consecutively, Jan. 1 to 11, 1874.
A spirit refusing to be misled by a suggestion.
A spirit earth-bound by love of money.
Influence of association, especially of locality.
Spirits who have communicated for a long period.
Child-spirits communicating: corroborative testimony from a second source.
Extremely minute evidence given by two methods.
A possible misconception guarded against.
General conclusions.
Personal immortality.
Personal recognition of and by friends.
Religious aspects.

APPENDIX I.—On the power of spirits to gain access to sources of information.

APPENDIX II.—On some phases of Mediumship bearing on Spirit-Identity.

APPENDIX III.—Cases of Spirit-Identity.
(a) Man crushed by steam-roller.
(b) Abraham Florentine.
(c) Charlotte Buckworth.

APPENDIX IV.—Evidence from spirit-photography.

APPENDIX V.—On some difficulties of inquirers into Spiritualism.

APPENDIX VI.—Spirit-Identity—Evidence of Dr. Stanhope Speer.

[ADVERTISEMENTS.]
W. H. HARRISON'S PUBLICATIONS.

Price 1s., cloth, red edges.

A CLERGYMAN ON SPIRITUALISM.

D. CLERICUS.

With a Dedication to the
REV. SIR WILLIAM DUNBAR, BART.;
And Some Thoughts for the Consideration of the Clergy,
By LISETTE MAKDOUGALL GREGORY.

This Booklet contains the experiences of a Clergyman, who prayerfully and continuously investigated Spiritualism for a long series of years.

Price 12s. 6d., cloth, red edges; demy 8vo.
Illustrated by various Full-page and Double-page Engravings.

TRANSCENDENTAL PHYSICS.

By F. ZÖLLNER,
Professor of Physical Astronomy at Leipsic University.
TRANSLATED BY C. C. MASSEY.

This illustrated work is unique in its character, and is one of the most remarkable and philosophical books ever published in connection with Spiritualism.

Price 5s., crown 8vo, cloth, red edges.

PSYCHIC FACTS.

Contains striking Selections from the Writings of Mr. WILLIAM CROOKES, F.R.S., Mr. C. F. VARLEY, F.R.S., Mr. A. R. WALLACE, F.R.G.S., the COMMITTEE of the DIALECTICAL SOCIETY, Professor HARE, of Philadelphia, Professor ZÖLLNER, Mr. SERJEANT COX, Captain R. F. BURTON, THE LORD LINDSAY, Dr. A. BUTLEROF, J. W. EDMONDS (Judge of the Supreme Court, New York), and other authors, demonstrating the reality of the phenomena of Spiritualism. The work contains some useful information for inquirers.

[ADVERTISEMENTS.]
W. H. HARRISON'S PUBLICATIONS.

Price 5s. 6d., crown 8vo, cloth, red edges.

THE FIRST VOLUME OF

SPIRITS BEFORE OUR EYES.

BY WILLIAM H. HARRISON.

This Book deals with the nature, characteristics, and philosophy of Apparitions, and how to reproduce experimentally some of the phenomena connected with them. It is also full of evidence of Spirit-Identity.

Price 2s. 6d., cloth, crown 8vo, red edges.

MESMERISM,
WITH HINTS FOR BEGINNERS.

BY JOHN JAMES
(*Formerly Captain, 90th Light Infantry*).

An excellent Text-Book by a writer who has had thirty years' experience in the subject.

W. H. HARRISON,
33 MUSEUM STREET, LONDON.

www.ingramcontent.com/pod-product-compliance
Lightning Source LLC
Chambersburg PA
CBHW030751230426
43667CB00007B/925